HOLLOW GLASS MICROSPHERES FOR PLASTICS, ELASTOMERS, AND ADHESIVES COMPOUNDS

PLASTICS DESIGN LIBRARY (PDL)
PDL HANDBOOK SERIES

Series Editor: Sina Ebnesajjad, PhD (sina@FluoroConsultants.com)
President, FluoroConsultants Group, LLC
Chadds Ford, PA, USA
www.FluoroConsultants.com

The **PDL Handbook Series** is aimed at a wide range of engineers and other professionals working in the plastics industry, and related sectors using plastics and adhesives.

PDL is a series of data books, reference works and practical guides covering plastics engineering, applications, processing, and manufacturing, and applied aspects of polymer science, elastomers and adhesives.

Recent titles in the series

Biopolymers: Processing and Products, Michael Niaounakis (ISBN: 9780323266987)
Biopolymers: Reuse, Recycling, and Disposal, Michael Niaounakis (ISBN: 9781455731459)
Carbon Nanotube Reinforced Composites, Marcio Loos (ISBN: 9781455731954)
Extrusion, 2e, John Wagner & Eldridge Mount (ISBN: 9781437734812)
Fluoroplastics, Volume 1, 2e, Sina Ebnesajjad (ISBN: 9781455731992)
Handbook of Biopolymers and Biodegradable Plastics, Sina Ebnesajjad (ISBN: 9781455728343)
Handbook of Molded Part Shrinkage and Warpage, Jerry Fischer (ISBN: 9781455725977)
Handbook of Polymer Applications in Medicine and Medical Devices, Kayvon Modjarrad & Sina Ebnesajjad (ISBN: 9780323228053)
Handbook of Thermoplastic Elastomers, Jiri G Drobny (ISBN: 9780323221368)
Handbook of Thermoset Plastics, 2e, Hanna Dodiuk & Sidney Goodman (ISBN: 9781455731077)
High Performance Polymers, 2e, Johannes Karl Fink (ISBN: 9780323312226)
Introduction to Fluoropolymers, Sina Ebnesajjad (ISBN: 9781455774425)
Ionizing Radiation and Polymers, Jiri G Drobny (ISBN: 9781455778812)
Manufacturing Flexible Packaging, Thomas Dunn (ISBN: 9780323264365)
Plastic Films in Food Packaging, Sina Ebnesajjad (ISBN: 9781455731121)
Plastics in Medical Devices, 2e, Vinny Sastri (ISBN: 9781455732012)
Polylactic Acid, Rahmat et al. (ISBN: 9781437744590)
Polyvinyl Fluoride, Sina Ebnesajjad (ISBN: 9781455778850)
Reactive Polymers, 2e, Johannes Karl Fink (ISBN: 9781455731497)
The Effect of Creep and Other Time Related Factors on Plastics and Elastomers, 3e, Laurence McKeen (ISBN: 9780323353137)
The Effect of Long Term Thermal Exposure on Plastics and Elastomers, Laurence McKeen (ISBN: 9780323221085)
The Effect of Sterilization on Plastics and Elastomers, 3e, Laurence McKeen (ISBN: 9781455725984)
The Effect of Temperature and Other Factors on Plastics and Elastomers, 3e, Laurence McKeen (ISBN: 9780323310161)
The Effect of UV Light and Weather on Plastics and Elastomers, 3e, Laurence McKeen (ISBN: 9781455728510)
Thermoforming of Single and Multilayer Laminates, Ali Ashter (ISBN: 9781455731725)
Thermoplastics and Thermoplastic Composites, 2e, Michel Biron (ISBN: 9781455778980)
Thermosets and Composites, 2e, Michel Biron (ISBN: 9781455731244)

To submit a new book proposal for the series, or place an order, please contact David Jackson, Acquisitions Editor
david.jackson@elsevier.com

HOLLOW GLASS MICROSPHERES FOR PLASTICS, ELASTOMERS, AND ADHESIVES COMPOUNDS

Edited by

Stephen E. Amos

Baris Yalcin

ELSEVIER

Amsterdam • Boston • Heidelberg • London
New York • Oxford • Paris • San Diego
San Francisco • Singapore • Sydney • Tokyo
William Andrew is an imprint of Elsevier

William Andrew is an imprint of Elsevier
The Boulevard, Langford Lane, Kidlington, Oxford, OX5 1GB, UK
225 Wyman Street, Waltham, MA 02451, USA

Library of Congress Cataloging-in-Publication Data
A catalog record for this book is available from the Library of Congress

British Library Cataloguing-in-Publication Data
A catalogue record for this book is available from the British Library

ISBN: 978-1-4557-7443-2

For information on all William Andrew publications
visit our website at http://store.elsevier.com/

Publisher: Matthew Deans
Acquisition Editor: David Jackson
Editorial Project Manager: Peter Gane
Production Project Manager: Nicky Carter
Designer: Maria Inês Cruz

Typeset by TNQ Books and Journals
www.tnq.co.in

Contents

Contributors

Stephen E. Amos 3M Advanced Materials Division, 3M Center, St. Paul, MN, USA

Warren Beck Retired, 3M Central Research, 3M Center, St. Paul, MN, USA

Cary Buller Applications Engineer, MYERS Engineering, Inc

Nikhil Gupta Composites Materials and Mechanics Laboratory, Department of Mechanical and Aerospace Engineering, New York University, Polytechnic School of Engineering, Brooklyn, NY, USA

Dinesh Pinisetty The California Maritime Academy, California State University, Vallejo, CA, USA

Stephanie Shira Applications Engineer, MYERS Engineering, Inc

Vasanth C. Shunmugasamy Composites Materials and Mechanics Laboratory, Department of Mechanical and Aerospace Engineering, New York University, Polytechnic School of Engineering, Brooklyn, NY, USA

Jean Tangeman 3M Corporate Research Materials Laboratory, 3M Center, St. Paul, MN, USA

Baris Yalcin 3M Advanced Materials Division, 3M Center, St. Paul, MN, USA

1 Introduction

Stephen E. Amos and Warren Beck

Since the dawn of the mankind, there has been a drive to develop lighter materials to enable transport and ease of use. After the industrial revolution and the subsequent development of plastics, there have been ongoing material substitutions from metal, glass, wood, and stone to plastics and composites of these materials to reduce weight. A logical next step in this material evolution was to reduce the weight of plastics. Various, naturally low in density, fillers were first tried with limited density modification capability. In addition, injection or creation of gas in the polymer during the article forming process was also developed and utilized in nonstructural applications such as packaging.

By the late-1930s, 3M Company was manufacturing solid glass beads made of scrap window glass. This product was sold to highway departments for reflective road paint. Various reformulating experiments were carried out to attempt to produce hollow glass microspheres (HGMs) but were limited in their success by low yields. By the 1950s, 3M was selling reflective sheeting to the French army, which was called "Army Cardboard"—2×2 sheets of reflective sheeting with low-refractive index glass beads. The sheets were optically designed to be retro reflective for light perpendicular to the sheet plane. These were mounted on the back of convoy vehicles to help prevent night-time accidents.

When the French government placed a large order for Army Cardboard, 3M made the material but it failed to meet the brightness requirements. A 3M scientist, Warren Beck was manager of the Bead Department and he undertook the task of determining why the feed material had failed. Like many scientific discoveries, what was perceived as failure was really a pathway to success for the development of a new product.

When Beck examined the out-of-spec material, he discovered clouds of microbubbles near the surface of the beads. He determined that storing the crushed glass feed particles, for a long period of time in humid weather, had created the conditions to form hollow bubbles. To correct the problem, he recommended crushing the glass and using it immediately. Case solved.

But Beck also knew of the preceding work within 3M attempting to develop such a hollow glass bead, and of earlier patented work by

Hollow Glass Microspheres for Plastics, Elastomers, and Adhesives Compounds
http://dx.doi.org/10.1016/B978-1-4557-7443-2.00001-3

Figure 1.1 Visual microscopic image of hollow glass microspheres made by 3M Company. *3M™ Glass Bubbles—Courtesy of 3M.*

Standard Oil of Ohio on a one step, melt and expand "microsphere" product and process based on either phenolic resin or sodium silicate glass [1,2]. After some experimentation, Beck discovered that it was possible to create hollow beads or "HGMs" as 3M would later call them with a two-stage melting and forming process. In 1963, he filed a patent application for creating these unique structures by carefully formulating glass frit, milling it to a specific particle size and particle size distribution, then reheating the particles to form single-wall hollow glass microsphere Figure 1.1 [3].

The Sohio patents were eventually sold to Philadelphia Quartz—PQ Corporation today. PQ currently makes HGMs of this type of glass. The phenolic based microspheres ended up being produced by Union Carbide Corporation.

There have been several other types of materials, discovered or developed over the years that also provide density modification for resin systems. Fly ash is the by-product of powdered coal-burning power plants. It is similar to impure clays in composition in that it is primarily aluminum silicate contaminated with iron, magnesium, calcium, and alkali metal oxides. As the coal particles burn, the ash, which can make up to 10% or more of the coal, fuses to form hollow microspheres. If composition and forming conditions are right up to several percent of the spheres that are produced may be hollow and low enough in density to float on water. This type of bubbled product was first recovered, floating in power-plant ash ponds, in England around 1970 and marketed as "cenospheres." The

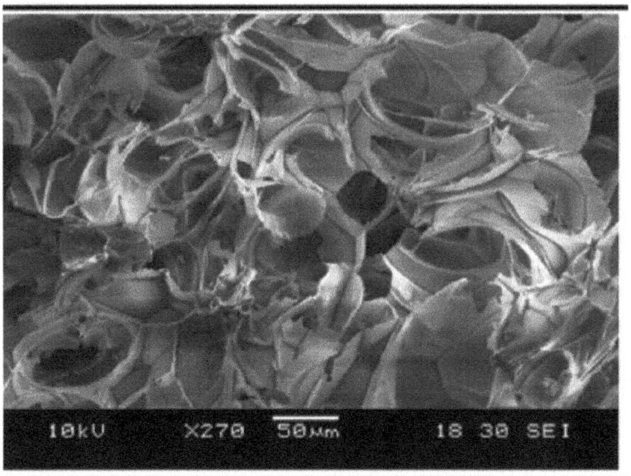

Figure 1.2 Expanded perlite. *Figure From Ref. [4] with Permission.*

density of cenospheres is generally around 0.7 g/cc and their strength is highly variable but usually around 3000 PSI due to imperfections in the sphere wall.

Perlite has been an item of commerce for a number of decades. When heated above the softening point (about 900° C), water internal to the perlite structure is liberated as steam and the material forms a porous, low density, multicellular material as shown in Figure 1.2.

When added to liquids or molten plastics, the pores can absorb resin to a degree, depending on the resin viscosity. Some pores are too small to be filled and remain as voids so the material can provide a small amount of density modification to a composite. Generally, perlite is severely degraded in high shear flow environments so it is typically used as a filler in thermoset systems, not in thermoplastics. But, it is primarily used in nonresin applications such as insulation fills in cryogenic liquid storage tanks. There have been various attempts to make fused single cell HGMs from expanding perlite [5].

Kanamite was a hollow ceramic particle, made from shale, having a particle diameter of 100–600μm. The density varied from 0.4 to 0.8 g/cc. The material was manufactured by the Kanium Corporation of Chicago in the 1960s. It is no longer manufactured today.

Early Application Examples

Various applications were promoted in the early patents for HGMs including the use in plastics, rubber, and other resinous materials for weight reduction. Other application areas of interest were thermal insulation, concrete, synthetic wood, gas storage and transport, and as a flow aid (the ball bearing effect).

One of the first successful applications for HGMs was in dry wall joint sealer. Normal, dense, plaster- or PVA-based joint sealer would shrink and crack requiring multiple applications. The HGM glass material has a very low coefficient of thermal expansion (CTE), preventing shrinkage. Also the wall joint material was very hard after curing and required a significant amount of work to sand to a smooth surface. Providing microvoid spaces improved the postcure processing properties allowing for quick sanding to a smooth surface. One benefit not immediately realized was that of light weighting. This prevents sagging of the compound on vertical surfaces.

An early, unexpected application was the use of HGMs in explosives. Prior to 1951, little was known about the explosive reaction between ammonium nitrate and fuel oil. But a disastrous explosion in Texas City that year resulted in studies leading to an understanding of the mechanism. The Dow Chemical Company, one of the blasting agent suppliers, discovered that this unreliability could be controlled by the incorporation of tiny air bubbles in the slurry. This was originally done by whipping the slurry, but there were problems in controlling the size and distribution of the voids. When the 3M HGMs became available, they were evaluated and eventually used for this job. Still today, cartridges of slurry blasting agents containing 1–2% of HGMs for stabilization have displaced dynamite in mining and construction applications.

Probably, the most obvious use for HGMs was as a functional filler for plastics to enhance properties and/or reduce costs. In the 1970s, the usage of HGMs was mostly for explosives and some for resin applications. Resin applications grew quickly and today these applications include dry wall joint sealer, autobody filler, grout, caulk, potting compound, plastisol, adhesives, sheet molding compound, bulk molding compound, marine applications, extruded and thermoset insulation, buoyancy modules, and thermoplastic injection molded parts for transportation and other applications. New applications combining light-weighting technologies [5] are being advocated in the plastics marketplace as many processors and end users grapple with questions of renewability, sustainability, CO_2 production and carbon taxes, Corporate Average Fuel Economy (CAFÉ Standards), fuel consumption, and release of greenhouse gases to the environment.

There are very few fillers or even additives that are employed in the plastic industry that are lower in density than a typical base resin. This makes these materials unique, and somewhat problematic to handle and

Figure 1.3 Equal weights of typical polymer fillers demonstrate the volume difference for low-density materials like HGMs. *With Permission from Reference 6—Courtesy of 3M.*

formulate. The advent of HGMs has made a couple of generations of formulating chemists and polymer scientists go back and relearn the difference between volume and weight filling in composite systems. As the old adage goes, a picture is worth a 1000 words—here is that picture shown in Figure 1.3.

When adding HGMs to resin systems, it is very important to determine the volume% of all of the materials present. Failure to do so may remove too much binder (resin) or dilute out reinforcing and other important fillers/additives. The second significant difference is the separation mechanism. The HGMs tend to float out on top of low-viscosity liquid systems instead of sinking to the bottom like other fillers and pigments.

There are compelling reasons to evaluate and use HGMs in plastic and rubber systems. Statistics on weight reduction and improvements in MPG talk about 1–2% MPG improvement with every 100 lbs of vehicle weight reduction. Extending that to the amount of fuel savings and CO_2 released over the lifetime of a vehicle can be very significant especially when larger, less fuel-efficient vehicles are the targets, or when long-lived vehicles such as aircraft are used almost 24/7 to maximize productivity. Removal of thermal mass can provide improved productivity to extrusion and molding processes by reducing cycle time. Spheres have the lowest surface area to volume of any particle and therefore cause less viscosity build while providing isotropic filling of resins (having an aspect ratio of 1 reduces anisotropy). The low CTE of the glass used in these HGM compositions can lower the CTE of a composite and improve fit and finish as well as reduce noise, vibration, and harshness. On the surface, these materials look like a one trick pony—providing weight reduction. But they do more—they provide a combination of properties for plastic composites that provide a compelling combination of benefits for processors and end users.

References

[1] F. Veatch, et al., US patent 2797201, June 25, 1957.
[2] F. Veatch, et al., US patent 3030215, April 17, 1962.
[3] W. R. Beck, et al., US patent 3365315, January 23, 1968.
[4] Materials Chemistry and Physics 115 (2009) 846–850.
[5] F. J. Brodmann, US patent 3961078, June 8, 1976.
[6] B. Yalcin, et al., 'Plug and Play' Weight Reduction Solution with Hollow Glass Microspheres, 3M.com Technical Paper, 2011.

2 Characterization

Baris Yalcin, Stephen E. Amos and Jean Tangeman

Some of the most common techniques used to characterize hollow glass microspheres (HGMs) and materials containing HGMs are described herein. The properties that are described here have been limited to those that are a direct result of the hollow nature of glass microspheres. These include density, strength, thermal, and electrical properties.

Density

Density is a major property for HGMs along with isostatic collapse strength. Suppliers of particulate fillers define density differently and comparing density values without knowing the exact method used for the measurement can mislead the user, especially if selecting fillers for formulation purposes. For example, one density definition that is occasionally used in the HGM industry is bulk density (also called apparent density) which is the weight of a sample in a unit volume container. It is easily determined by filling a container of known volume (e.g., 1 L or 1000 cc) with HGMs to the very top and measuring the weight of the HGMs (e.g., in grams) and dividing the weight by volume (1000 cc). The problem with using bulk density is that the volume of the container contains not only the volume of the HGMs themselves, but also the air spaces between the microspheres. For practical purposes, bulk density is only useful for determining the size of the container needed to package or store the filler materials.

Another method of density measurement utilizes settled (tapped) volume of powders. In this test, 50 ± 0.2 g of HGMs are placed in a 100 ml graduated cylinder, and then tapped 3000 times. The tap density of the HGMs is measured by mass divided by the volume to which they settle after the vibration treatment [1]. However, settled (tapped) density method, similar to bulk density, does not exclude the air spaces between the HGMs and cannot be used for formulation purposes.

For formulating purposes, the true density is the most appropriate density parameter. This is the weight of the sample for the volume occupied only by the particles. It does not include the air spaces between

Hollow Glass Microspheres for Plastics, Elastomers, and Adhesives Compounds
http://dx.doi.org/10.1016/B978-1-4557-7443-2.00002-5

Figure 2.1 Bulk density and true density demonstration. *Courtesy of 3M.*

the microspheres. Ratio of bulk density to true particle density gives the packing factor which averages close to 0.6 for most spherical objects.

Figure 2.1 illustrates bulk density calculation by large solid glass spheres in a beaker. It simulates the way in which HGMs lie together in a container but are not packed against themselves or against the walls of the container. In this example, 275 g of solid glass spheres occupy 200 mL in the beaker which is a bulk density of 1.38 g/cc (275 g/200 ml). We know from literature that the true density of solid glass is approximately 2.5 g/cm^3, which means that if there was one solid piece of glass in that container, the bulk density and the true density would be the same. The difference in volume is approximately 90 cm^3 of air in between the particles. When 90 cm^3 of fluid is added to that same beaker with the glass spheres in it, it would still have the same 200 cm^3 of total bulk volume. This is exactly what happens when HGMs or other fillers are used in liquid resin system. The liquid displaces the air and completely fills the spaces around the particles.

The true density is the same 275 g of glass spheres divided by 110 cm^3, which is the volume occupied only by the glass spheres. The true density is 2.5 g/cm^3.

True density is more useful for formulation work because it is used for calculating what is referred to as the bulking factor. The bulking factor is the amount of liquid that is displaced by a given weight of HGMs. This number is the same as the reciprocal of the true density. For HGMs that have a true density of 0.125 g/cm^3, the reciprocal is 8 cm^3/g. With this number the predicted density of mixtures can be calculated. For instance, if 10 g of a resin has a density of 1 g/cm^3 and you add 1 g of HGMs (0.125 g/cc), the total weight is 11 g. Ten grams of resin at 1 g/cm^3 occupies 10 cm^3. The 1 g of HGM of 0.125 g/cc occupies 8 cm^3. The total mixture volume is 18 cm^3. Eleven grams of resin divided by 18 cm^3 calculates to a final predicted density of 0.61 g/cm^3.

Another challenge with bulk density is that it is not a reproducible number. In this example with the large glass spheres, 275 g usually fits into the 200 cm^3 container. For some of the mineral fillers that do not pack well together, it is very difficult to get the same volume reading every time they are poured into a container. Data sheets for minerals sometimes list two bulk density numbers. One number is for loosely packed material and the other number is for tightly packed material. It is not easy to define "loosely" or "tightly" packed.

True density can be measured with a liquid displacement method, just as we have shown above, with the large *solid* glass spheres (solid beads in Figure 2.1). However, for low-density HGMs, it is difficult to use the liquid displacement method due to floatation. Fortunately, there are scientific instruments such as gas comparison pycnometers that will perform these measurements [2–4].

True Density Using Gas Pycnometer

True density of HGMs can most conveniently be measured using gas pycnometry, which is a common analytical technique that uses a gas displacement method to measure volume accurately. In this method, the sample is sealed in the instrument compartment of known volume (V_c), the appropriate inert gas (e.g., helium, nitrogen) is admitted, and then expanded into another precision internal volume (V_r). The pressure before (P_1) and after expansion (P_2) is measured and used to compute the sample volume (V_s) as shown in Figure 2.2 using the ideal gas law. Dividing this volume into the sample weight gives the density. In Figure 2.2, inert gas (helium or nitrogen) which fills in between the spaces of the hollow spheres completely is depicted by the green color. For ease of the understanding of the calculations, the total true volume of the HGMs (V_s) is represented by a large irregular shape shown in Figure 2.2.

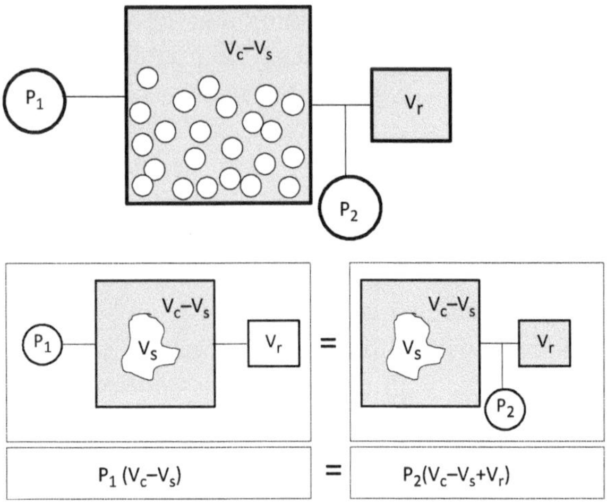

Figure 2.2 Working principle of a gas pycnometer for measuring true density.

$$
\begin{aligned}
P_1V_1 &= P_2V_2 @ \text{ Constant T and Mass (ideal gas law)} \\
P_1(V_c - V_s) &= P_2(V_c - V_s + V_r) \\
V_s &= V_c + (V_r/(1 - P_1/P_2)) \\
\rho_s &= M_s/V_s
\end{aligned}
$$

where

V_S: sample volume (true volume of the hollow spheres);

V_C: volume of the empty sample chamber;

V_r: volume of the reference volume;

P_1: first applied pressure (i.e., in the sample chamber only, approximately around 19.4 psi);

P_2: second (lower) pressure after expansion of the gas into the combined volumes of sample and reference chamber;

M_S: input mass of the hollow spheres; and

ρ_S: density of the hollow spheres.

The true density of HGMs is useful not only to determine the composite density in a matrix resin as shown in Eqn (2.1) but also to determine the volume percentage of HGMs broken during the incorporation process into the resin and/or post processing of HGM containing resins. When making any type of product that includes HGMs one concern is whether the manufacturing processes rupture any of the HGMs.

$$\rho_{\text{resin/HGM composite}} = \left[\frac{1}{\left(\frac{Wf_{\text{HGM}}}{d_{\text{true HGM}}}\right) + \left(\frac{Wf_{\text{resin}}}{\rho_{\text{resin}}}\right)}\right] \tag{2.1}$$

where

Wf_{HGM}: weight fraction of HGM in the formula expressed as a decimal (e.g., 10% would be 0.1);

Wf_{resin}: weight fraction of resin in the formula expressed as a decimal (e.g., 90% would be 0.9); and

$d_{\text{true HGM}}$: true density of HGM.

As HGMs break, their average density approaches that of solid glass used to construct the wall. At 100% breakage, the density of HGMs reaches 2.54 g/cc, the density of solid glass. In order to determine the extent of HGM breakage reliably, the HGMs in the composite resin need to be extracted from the matrix resin. This is readily accomplished, for instance, in plastic resin matrices by exposing the pellets or molded parts, to high temperature (e.g., 550 °C for 8 h in a kiln) to remove the plastic phase. The residual inorganic ash would merely consist of HGMs, the true density of which can be calculated using a gas pycnometer. Equation (2.2) can then be used to determine volume percentage of HGM broken in process;

$$\%\text{Vol of HGM broken in process} = \left[\frac{\left(\frac{1}{(d_{\text{true HGM-input}})}\right) - \left(\frac{1}{(d_{\text{true HGM-residue}})}\right)}{\left(\frac{1}{(d_{\text{true HGM-input}})}\right) - \left(\frac{1}{d_{\text{true glass}}\ (2.54)}\right)}\right] \times 100 \tag{2.2}$$

Example: What is the volume percentage of HGM broken after extrusion in a polymer, if the input density of HGM used is 0.46 g/cc and the density of residual ash after burn off process is measured to be 0.52 g/cc.

Answer: Using Eqn (2.2),

$$\%\text{Vol of HGM broken in process} = \left[\frac{\left(\frac{1}{(0.46)}\right) - \left(\frac{1}{(0.52)}\right)}{\left(\frac{1}{(0.46)}\right) - \left(\frac{1}{(2.54)}\right)}\right] \times 100 = 14\%$$

If the formulation has filler or other inorganic additives (e.g., glass fibers) in addition to HGMs, the residual ash after removing the matrix resin would comprise HGMs and the other inorganics. In such a case, residual inorganic density ($\rho_{\text{glass fiber \& bubble mix}}$) can be determined using the gas pycnometer method described above. HGM density in residue ($\rho_{\text{HGM in residue}}$) can then be extracted from the Eqn (2.3) knowing the other filler density (e.g., 2.5 g/cc for glass fiber or 2.68 g/cc $CaCO_3$) and the density of the inorganic residue $\rho_{\text{other filler \& bubble mix}}$.

$$\rho_{\text{other filler \& HGM mixture}} = \frac{1}{\frac{Wf_{\text{other filler}}}{\rho_{\text{other filler}}} + \frac{Wf_{\text{HGM}}}{\rho_{\text{HGM in residue}}}} \qquad (2.3)$$

Once the HGM density in residue ($\rho_{\text{HGM in residue}}$) is determined, one can again use Equation (2.2) to calculate volume loss.

Example: What is the volume percentage of HGM broken in a polymer containing 10 wt% HGM (input density 0.46 g/cc) and 20 wt% glass fiber (input density 2.5 g/cc), if the density of residual ash after burn off process is measured to be 1.38 g/cc.

Answer: A polymer containing 10 wt% HGM and 20 wt% glass fiber would result in 33.3 wt% HGM and 66.7 wt% glass fiber after removing the polymer phase.

$$1.38 = \frac{1}{\frac{0.667}{2.50} + \frac{0.333}{\rho_{\text{HGM in residue}}}} \qquad \text{(from Eqn 2.3)}$$

$$\rho_{\text{HGM in residue}} = 0.727 \text{ g/cc}$$

$$\%\text{Vol of HGM broken in process}$$

$$= \left[\frac{\left(\frac{1}{(0.46)}\right) - \left(\frac{1}{(0.727)}\right)}{\left(\frac{1}{(0.46)}\right) - \left(\frac{1}{(2.54)}\right)}\right] \times 100 = 44.8\% \qquad \text{(from Eqn 2.2)}$$

If the HGMs cannot be extracted from the matrix resin reliably, one can first calculate the final density of the HGM containing part. The degree of HGM breakage can then be calculated with the following Eqn (2.4):

$$\%\text{Vol of HGM broken} = \frac{(\rho_{HGM}) \times (\rho_2 - \rho_1) \times 100}{W_{f-HGM} \times \rho_1 \times \rho_2} \qquad (2.4)$$

where:

ρ_{HGM} = density of input HGMs in g/cc;

ρ_1 = initial density in g/cc (composite part density if no HGMs were broken);

ρ_2 = final density in g/cc (actual composite part true density);

W_{f-HGM} = weight fraction of HGM in the formula expressed as a decimal.

Example: A mixture of 90 parts by weight polyester resin (density = 1.04 g/cc) and 10 parts by weight HGM (density = 0.20 g/cc) are mixed. After the two are mixed together, the density is measured and the result is 0.78 g/cc. What is the volume percentage of HGM broken?

Answer: If there was no HGM breakage, the final part density would be,

$$\rho_{\text{resin/HGM composite}} = \left[\frac{1}{\left(\frac{0.1}{0.2}\right) + \left(\frac{0.9}{1.04}\right)}\right] = 0.73\ \text{g/cc}$$

Knowing that the actual part density is 0.78 g/cc, one can calculate volume percentage of HGM broken spheres.

$$\%\text{Vol of HGM broken} = \frac{(0.2) \times (0.78 - 0.73) \times 100}{0.1 \times 0.73 \times 0.78} = 17.6\%$$

It is important to note that in many cases a significant amount of air may be mixed into a compound along with the HGMs. If this happens, the density will be lower than theoretical. However, if some HGMs were broken, the increase in density may be masked by the air.

Bulk Isostatic Compression of HGMs

There are various methods of measuring HGM compressive properties. One way to test compressive properties (collapse strength) of a batch of HGMs is in isostatic form which was described in detail by Carlisle et al. [5]. Bulk isostatic compression measurement techniques allow for rapid characterization of the compressive strength of producing lots of HGMs in an industrial setting. The test method used to characterize compressive properties of hollow spheres was at one time, the American Society for Testing and Materials (ASTM) Standard, bearing designation D 3102-78 [6]. Today, it is used by HGM manufacturers with modifications and exceptions. The basic premise of the test is to expose a batch of spheres to a high isostatic pressure, during which some of the spheres will collapse. The percent of spheres that collapse (or survive) is measured, and the spheres are classified according to the isostatic pressure at which a given percent of the spheres tested collapse (or survive). The spheres are placed in a rubber container filled (typically an inflatable rubber balloon) with glycerin inside a pressure chamber. Glycerin is used because it is one of the least compressible liquid.

Once the sample is in the pressure chamber, it is pressurized to a predetermined pressure while recording both pressure and volume. This process is repeated to get a second set of pressure–volume data, and both data sets are plotted as pressure versus volume curves. The initial pressurization is known as the collapse curve, and the second pressurization provides the system compression curve. Figure 2.3 provides an example of the two curves, which are used to determine the percentage of collapsed HGMs in the test.

Percent volume survival ($V_{survival}$) calculations shown in Eqn (2.5) are reported in the reference in detail.

$$\%V_{survival}(P) = 1 - \left[100 \times (V_{total(P=0)} - (V_1(P) - V_2(P))/(V_{ap} - V_m)\right] \tag{2.5}$$

where:

$V_1 = V_{compression}$;

$V_2 = V_{collapse}$;

$V_T = V_{total\ (P=0)}$: total volume collapse calculated by the difference in the collapse and compression curve values at zero pressure;

$V_1(P)$ and $V_2(P)$ are the volume values of the compression and collapse curves, respectively, at the pressure of interest;

V_{ap}: volume occupied by the HGMs; and

V_m: volume of the wall material.

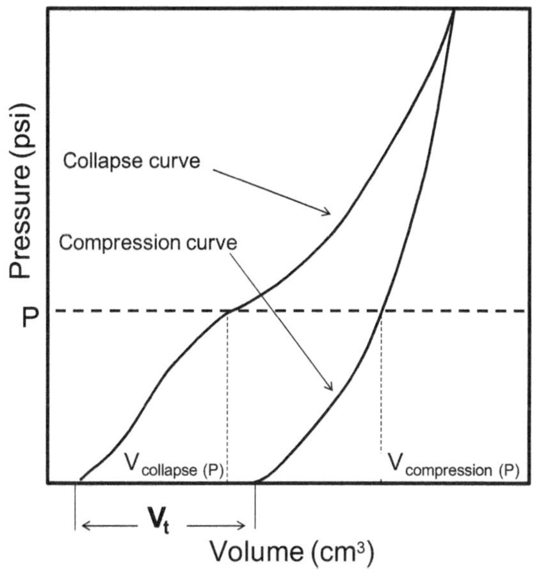

Figure 2.3 Bulk isostatic compression test curves. *With permission from Ref. [5].*

Uniaxial Compression Testing of Individual HGMs

One of the more recent developments in strength testing of glass HGMs involves the use of a modified nanoindenter for uniaxial compression testing of individual hollow spheres. For this procedure the standard Berkovich tip used in nanoindentation is replaced with a 90-μm-diameter cylindrical sapphire indenter, and a flat polished aluminum substrate is placed into the sample stage to function as a lower compression platen. Proper alignment is critical to uniaxial testing of individual HGMs and is accomplished using an optical microscope, which is part of the nanoindenter, along with a position calibration technique. In addition to revealing the ultimate compressive load of the HGMs, data collected using this technique provide several other useful engineering parameters. Load and displacement at failure (P_f and δ_f, respectively) are obtained directly from most load–displacement curves. The slope of the initial linear loading segment gives pseudostiffness (k), and the work of fracture (W_f) is defined as the area under the load–displacement curve up to the fracture point portion of the curve. The compressive failure strain (ε_f) is calculated as the quotient of the displacement at failure and the vertical diameter of the HGM in the loading direction [7,8].

Interestingly, and in contrast to results of bulk isostatic compression measurements, results on the uniaxial compression testing of individual HGMs have shown that the work of fracture and the failure load is positively correlated with HGM diameter. In other words, larger and presumably thinner-walled bubbles break at higher failure loads than their smaller, thicker-walled counterparts. Reasons for this may include (1) a thin-walled HGM is statistically less likely compared to a thicker-walled bubble to contain a flaw of critical size that would initiate a crack under a given load, and (2) larger spheres have a smaller radius of curvature relative to smaller spheres, so as the uniaxial compression test proceeds, the contact area between the larger bubble and the platen increases, lowering the stress on the sample by distributing a comparable load over a larger area [7,9,10].

Thermal Characterization

HGMs typically exhibit glass transition (softening) (T_g) temperatures of around 600 °C or less. At temperatures just above T_g, the material is thermodynamically a liquid that is metastable and of high viscosity ($\sim 10^7 - 10^{13}$ dPa × s). Differential thermal analysis (DTA) is a common method to determine glass transition, crystallization, and melting temperatures of glass materials. With this technique, a sample and a reference pan are heated from room temperature to as high as 1600 °C at a constant scan rate (typically 10°–20°/min). The differential heat flow is then plotted as a function of temperature and reveals the temperature and magnitude of endothermic or exothermic phase transitions or other events. Differential scanning calorimetry (DSC) is a related technique which combines three DTA scans (baseline, standard of known heat capacity, and sample) to enable quantification of the thermodynamic quantity, C_P, which is the heat capacity at constant pressure.

Thermal Conductivity

Thermal conductivity of HGMs is a function of their hollow volume and wall thickness. There have been several references where thermal conductivity of HGMs and HGM-filled composites were discussed [7,10–14].

Measurement of the thermal conductivity of individual HGMs is not a trivial task. On the other hand, measurement of a bulk quantity of HGMs is misleading due to the air spaces in between them. Therefore, it is most convenient to estimate the thermal conductivity of HGMs from theoretical models.

If one considers a HGM to be a composite of glass and interior space then there are several mathematical models that may be used to calculate the thermal conductivity. The simplest is a series model shown in Eqn (2.6).

$$k_{HGM} = V_{glass}\,k_{glass} + V_{interior}\,k_{interior} \tag{2.6}$$

where

k_{HGM} = thermal conductivity of the HGM;

k_{glass} = thermal conductivity of the glass;

$k_{interior}$ = thermal conductivity of the interior;

V_{glass} = volume fraction of the glass; and

$V_{interior}$ = volume fraction of the interior.

The series model is overly simplistic. Empirical data on the thermal conductivity of various HGMs typically are significantly lower than predicted by this model. A more sophisticated model would be to use a flux law model, the simplest of which is the Maxwell equation [15–17]. Thermal conductivity of HGMs can be predicted using the Maxwell flux law model shown in Eqn (2.7).

Maxwell equation:

$$\begin{aligned} k_{HGM}(T) = {} & k_{glass}(T)\Big[k_{interior}(T) + 2k_{glass}(T) \\ & - 2V_g\big(k_{glass}(T) - k_{interior}(T)\big)\Big] \Big/ \Big[k_{interior}(T) + 2k_{glass}(T) \\ & + V_{inteiror}\big(k_{glass}(T) - k_{interior}(T)\big)\Big] \end{aligned} \tag{2.7}$$

where

k_{GB} (T) = thermal conductivity of the HGM at temperature T;

k_{glass} (T) = thermal conductivity of the glass at temperature T;

$k_{interior}$ (T) = thermal conductivity of the interior medium at temperature T;

V_{glass} = volume fraction of the glass and

$V_{interior}$ = volume fraction of the interior medium.

Table 2.1 Thermal Conductivity of Glass and Interior Medium As a Function of Temperature

T (°C)	$k_{interior}$	k_{glass}
−100	–	0.836
−40	0.0068	–
0	0.0080	1.046
50	0.0095	–
100	–	1.171

Thermal conductivity equations for the glass material and interior medium as a function of temperature are derived empirically as shown in Table 2.1.

$$k_{interior}(T) = [0.0000293 \times T\ (^{\circ}C)] + 0.00799$$
$$k_{glass}(T) = \left[0.0017 \times T\ (^{\circ}C)\right] + 1.0177$$

Using the Maxwell model the thermal conductivity of various HGMs can be plotted as a function of density as shown in Figure 2.4. As expected, there is quite a linear relationship between density and thermal conductivity.

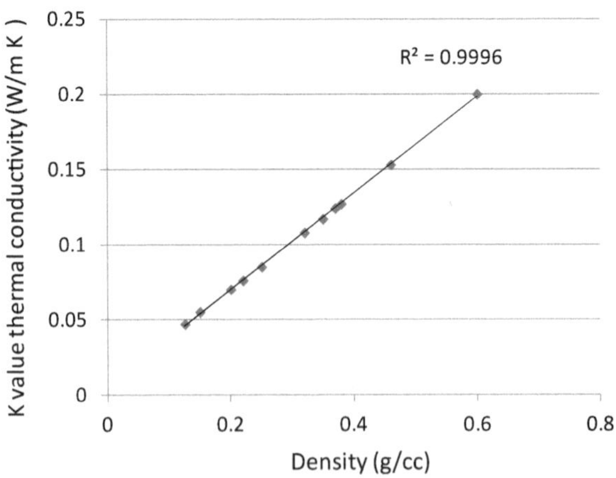

Figure 2.4 Calculated thermal conductivity as a function of density.

Thermal conductivity of HGM containing resins can be approximated by the Nielsen model [18] (Eqn (2.8)) which takes into account both the shape of HGMs (spherical) and the packing factor:

$$k_c = k_m(1 + ABV_{GB})/(1 - B\Psi V_{GB}) \tag{2.8}$$

where

$$B = [(k_{GB}/k_m) - 1)]/[(k_{GB}/k_m) + A)]$$
$$\Psi = 1 + \left[(1 - \Phi)/\Phi^2\right]V_{GB}$$

and

k_c = thermal conductivity of the composite;

k_m = thermal conductivity of the matrix;

k_{GB} = thermal conductivity of the HGM;

V_{GB} = volume fraction of the HGMs;

$A = 1.5$ for spherical particles; and

Φ = packing factor for HGMs $\cong 0.63$.

Table 2.2 compares experimentally determined and theoretically calculated (Nielsen model) thermal conductivity K-values (W/mK) of polypropylene containing 30, 40, and 50 wt% HGM (0.46 g/cc) at temperatures of 5, 30, and 60 °C. Experimentally measured values were determined using a heat flow meter according to ASTM C518 [19]. K-value samples were cut out in the form of a rectangular plaque (3.12 × 50 × 40 mm) from injection molded parts. The Lasercomp Fox 50 flow meter with WinTherm50 Version 2.30.01 software program was used with a Perspex calibration standard per ASTM-C518. One can see that the Nielsen model coupled with Maxwell can closely estimate the thermal conductivity of PP containing HGMs as a function of HGM loading and temperature (Figure 2.5).

Electrical Properties and Dielectric Properties

Similar to thermal conductivity, void volume of HGMs render some useful dielectric properties to the materials that they are in [20]. HGM-filled materials are typically characterized by low dielectric

Table 2.2 Measured and Calculated Thermal Conductivity K-value (W/mK) of PP Samples Containing GBs

	W/mK at 5 °C		W/mK at 30 °C		W/mK at 60 °C	
	Measured	**Theoretical**	**Measured**	**Theoretical**	**Measured**	**Theoretical**
30 wt% HGM (0.46 g/cc) in PP	0.1674	0.1685	0.1719	0.1719	0.1769	0.1759
40 wt% HGM (0.46 g/cc) in PP	0.1601	0.1597	0.1643	0.1639	0.1691	0.1689
50 wt% HGM (0.46 g/cc) in PP	0.1550	0.1536	0.1598	0.1584	0.1648	0.1640

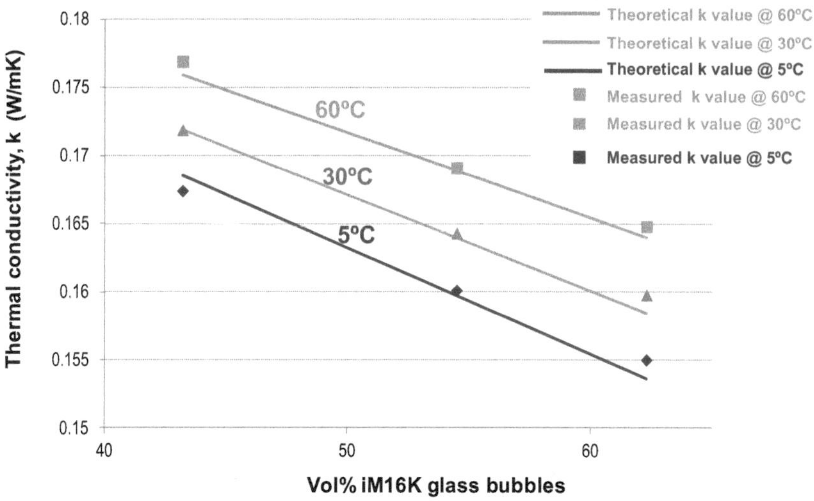

Figure 2.5 Comparison of measured (markers) and theoretical (lines) thermal conductivity at 5, 30, and 60 °C. Measured values were obtained using Fox 50 parallel plate assembly and theoretical thermal conductivity values were obtained using the Nielsen equation.

constants and low loss tangents. This enables their use, for example, in radomes, circuit substrates, packaging materials, and other products where lightweight materials of desired electrical properties are required. Low dielectric constants and low loss tangents increase the velocity of signal propagation while reducing signal attenuation avoiding major losses of radiation energy, especially as the working frequency increases [21–23].

Dielectric constant is the ratio of the strength of an electric field in a vacuum to that in the material for the same distribution of charge. It may also be defined and measured as the ratio of the capacitance, C, of an electrical condenser filled with the dielectric material to the capacitance C_0 of the evacuated condenser:

$$\varepsilon = C/C_0 \tag{2.9}$$

A vacuum has a dielectric constant of 1.00 and has the lowest dielectric constant. Dielectric constant of all materials is therefore greater than 1. The dielectric constant of a low-thermal-expansion borosilicate glass (such as that used in laboratory glassware), for example, is about 5 and soda lime window glass is about 7.5. However, if the glass is in the form of a hollow sphere, the dielectric constant of the hollow sphere is intermediate between that of air and glass, based on the void volume and glass volume of the sphere.

Dielectric constant of HGMs can be estimated using a simple rule of mixture model:

$$\varepsilon_{HGM} = V_{glass}\,\varepsilon_{glass} + V_{VOID}\,\varepsilon_{VOID} \quad (2.10)$$

where:

ε_{HGM} = dielectric constant of HGMs;

ε_{glass} = dielectric constant of glass;

ε_{VOID} = dielectric constant of void;

V_{VOID} = volume fraction of the void = true density/solid glass density; and

V_{glass} = volume fraction of the glass = $1 - V_{VOID}$.

Table 2.3 shows calculated dielectric constant values of HGMs of various densities using Eqn (2.10) assuming dielectric constant of glass and void to be 5 and 1, respectively.

Dielectric constant for the HGMs can be measured in a liquid medium of known dielectric constant such as silicon oil.

$$\varepsilon_{HGM} = (\varepsilon_i - \varepsilon_o(1 - f))/f \quad (2.11)$$

where

ε_{HGM} = dielectric constant of HGMs;

ε_i = instrument reading (i.e., dielectric constant of medium and HGM mix);

ε_o = dielectric constant of known medium, for example, silicon oil; and

f = volume fraction of HGMs.

Table 2.3 Dielectric Constant Values of HGMs

True Density (g/cc)	Volume Fraction of Glass	Volume Fraction of Void	Calculated Dielectric Constants (100 MHz)
0.125	0.050	0.950	1.20
0.37	0.148	0.852	1.59
0.46	0.184	0.816	1.74
0.60	0.240	0.760	1.96

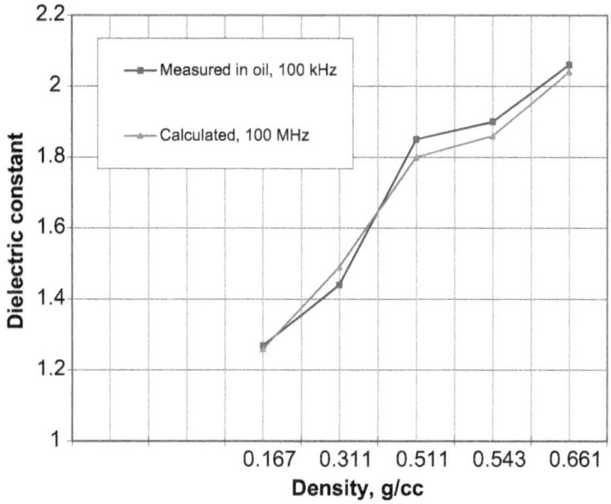

Figure 2.6 Calculated (Eqn (2.11)) and measured dielectric constants.

Figure 2.6 compares room temperature calculated (Eqn (2.11)) and measured dielectric constants using a general radio capacitance bridge at 100 kHz for HGMs with a range of densities. Although the calculations were made using a dielectric constant value for glass at 100 MHz, the values can be used for rough estimates at other frequencies too since the range for dielectric constant of glass with frequency typically lies in 4.5−5 at room temperature.

Dielectric constants can also be estimated in dry packed form. In this mode, packing factor (true density/bulk density) is used along with the measured dielectric constant of packed HGM (HGM plus air) to derive the dielectric constant of HGMs without air in between them.

$$\varepsilon_{HGM} = (\varepsilon_i - 1.00 + p_f)/p_f \qquad (2.12)$$

where:

ε_{HGM} = dielectric constant of HGMs;

ε_i = instrument reading (i.e., dielectric constant of HGM plus air); and

p_f = packing factor = true density/bulk density.

Vector network analyzers with shielded, open coaxial line probes were also used to measure the dry powder dielectric constant of HGMs. The shielded, open coaxial line is filled with HGMs and tapped to promote packing. Dielectric constant and loss are calculated directly from the complex reflection coefficient. Dielectric constant and loss for the HGMs are then calculated from a volume relationship, knowing the volume of HGMs and air in the probe. The following equations are used:

$$\varepsilon_{HGM} = \varepsilon_r(1 - p_f)/p_f \tag{2.13}$$

$$\text{Tan delta} = \varepsilon_i/(\varepsilon_r - (1 - p_f)) \tag{2.14}$$

where:

ε_{HGM} = dielectric constant of HGMs;

ε_r = the real part of the dielectric constant;

ε_i = the imaginary part of the dielectric constant; and

p_f = packing factor = true density/bulk density.

Dielectric constant and loss measurement of HGM containing resins can readily be measured using an impedance analyzer with applied AC voltages over a frequency range, for example, 1 kHz to 1 MHz. Yung et al. measured dielectric constant of HGM-filled epoxy composites on an Agilent-4294A impedance analyzer with an applied AC voltage of 500 mV over the frequency range 1 kHz to 1 MHz, with HGM filler content ranging from 0 to 51.3 vol% [24]. They used disc-shaped samples with a diameter of 12.6 mm, and a thickness of 1.5−4.0 mm and both sides of the samples were coated with silver paste. The D_k was calculated from capacitance by $D_k = Ct/e_0A$, where t was the thickness of the discs, e_0 the vacuum dielectric constant, and A the disc area. They found that both the dielectric constant (D_k) and dielectric loss (D_f) of the composites decreased simultaneously with increasing HGM content as shown in Figure 2.7, which was critical for the provision of superior high-frequency device performance.

Figure 2.8 shows the typical curves of the D_k and D_f versus the frequency from 10^3 to 10^6 Hz for 11.6 vol% HGM-filled epoxy. As shown in Figure 2.8, the D_k decreases but the D_f increases with increasing frequency for all the samples. The dielectric behavior involves different polarizations, and the polarization rate is dependent on temperature and frequency. At low frequencies, the polarization will have more time to complete compared with that at

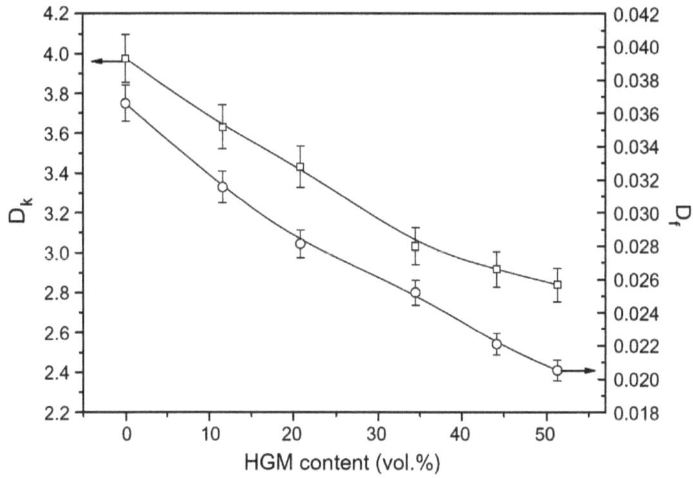

Figure 2.7 D_k and D_f as a function of hollow glass microsphere (HGM) content at the frequency of 1 MHz. *With permission from Ref. [24].*

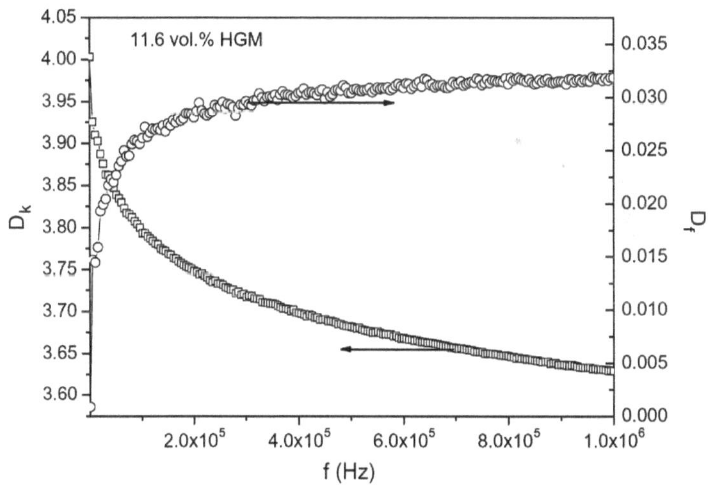

Figure 2.8 D_k and D_f of 11.6 vol% hollow glass microspheres (HGM) filled epoxy as a function of frequency. *With permission from Ref. [24].*

high frequencies. Thus the degree of polarization of material is high and the dissipation of polarization is low at low frequencies. This reveals that the D_k decreases but the D_f increases with increasing frequency.

Microscopic Imaging of HGMs

Scanning electron microscopy (SEM) is an invaluable tool for analyzing HGMs or HGM containing resins to obtain information such as particle size and distribution, morphology, and bubble integrity. Compositional mode of SEM imaging relies on density differences and can readily contrast glass (white) against polymer (gray) as shown in Figure 2.9.

Topography mode of imaging is also very useful since it identifies not only the HGMs but also HGM pockets if they are removed from their locations during sample preparation under freeze fracture. This gives good positional information about the HGMs where one can take several measurements between or on each sphere as shown in Figure 2.10. Liquid nitrogen freeze fracture is a convenient method of sample preparation to image cross-sections of parts containing HGMs. Cutting samples with scissors or other sharp objects to create an image plane would destroy HGMs and it is not recommended.

Sometimes, spheres are broken on purpose to provide wall thickness data as shown in Figure 2.11. However, for accurate measurement of wall thickness, the broken wall must be perpendicular to the microscope axis,

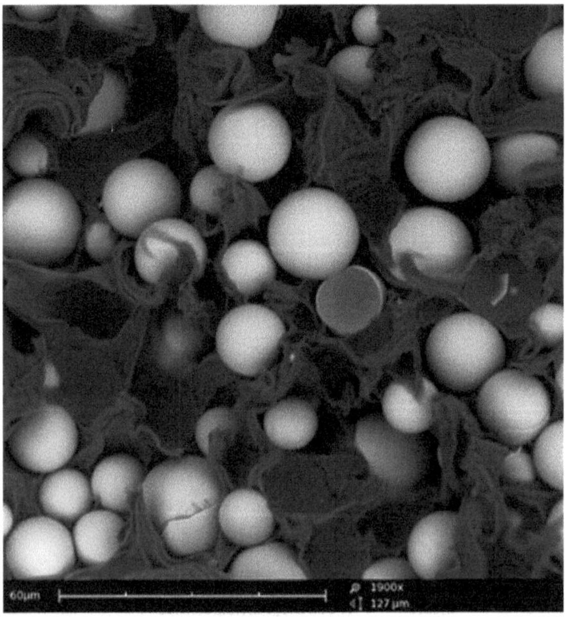

Figure 2.9 Compositional mode of scanning electron microscopy imaging of hollow glass microspheres. *Courtesy of 3M.*

Figure 2.10 Topography mode of scanning electron microscopy imaging of hollow glass microspheres (HGMs). *Courtesy of 3M.*

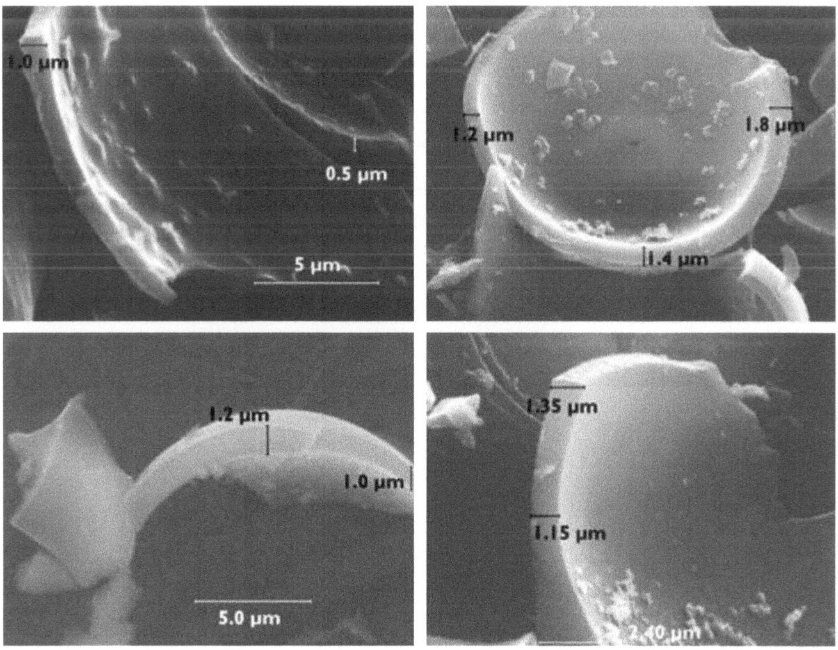

Figure 2.11 Wall thickness planes of broken spheres. *With permission from Ref. [7].*

or the angle of inclination known, for correction. Wall thickness measurement of spheres using SEM is specific to each sphere imaged and does not give an average value. An average wall thickness of a certain grade of HGM is typically estimated by deriving the inner radius of the sphere from known values of true density of the hollow spheres, absolute density of solid glass (2.54 g/cc), and D50 outer average particle diameter of the spheres using the following Eqn (2.15).

$$R_i = \sqrt[3]{R_o^3 - \frac{R_o^3 \times d}{2.54}} \tag{2.15}$$

where

R_i: inner radius of the sphere;

R_o: outer radius of the sphere ($R_o = D50/2$);

D50: outer average particle diameter of the spheres; and

d: true density of HGM.

Solid glass density: 2.54 g/cc

Wall thickness can then be calculated from known values of R_o and R_i

$$\text{wall thickness} = R_o - R_i \tag{2.16}$$

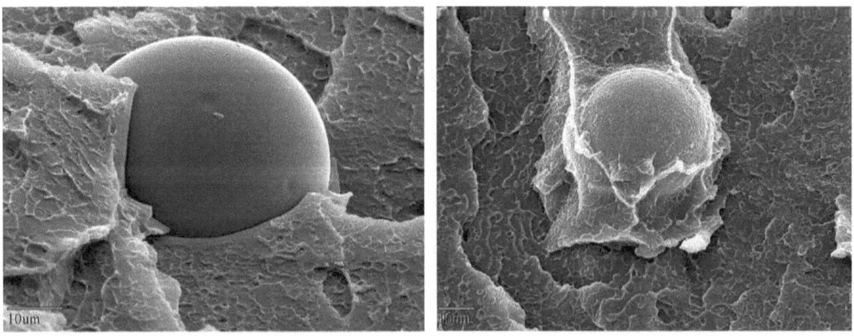

Untreated bubble Treated bubble

Figure 2.12 Scanning electron microscopy image showing level of bonding of the hollow glass microsphere to the matrix resin: (a) untreated bubble and (b) treated bubble. *3M Web site.*

SEM can also be used to assess bonding of HGMs to resins (Figure 2.12). HGMs with poor bonding show clean break interfaces whereas HGMs bonded strongly show well interlaced structures with the surrounding resin.

Conventional optical microscopy can also provide valuable and rapid information on HGM samples. Samples for optical microscopy analysis can be prepared by mounting in epoxy and then grinding and polishing to give planar, 2D slices through the surface of the HGMs. Sample preparation can also be as simple as sprinkling a monolayer of bubbles on an adhesive surface, for example, Scotch® tape. The sample is then placed on the stage and observed under the microscope using reflected, incident, or transmitted light.

For plastics containing HGMs, optical microscopy is useful for imaging the overall distribution of HGMs in the plastic part. Thermoplastics such as polyolefins and polyamides can be microtomed into thin slices (25–30 μm) and imaged in unpolarized or polarized mode of optical microscopy. For instance, in order to determine the HGM distribution from skin to core of the injection molded part, 25-μm-thick slices were taken from the injection molded samples in the normal–transverse (ND–TD) plane as well as normal-flow plane as shown in Figure 2.13. The sliced samples were sandwiched between glass slides and images were taken in transmission unpolarized mode with a Zeiss Axioplan optical microscope.

The optical image in Figure 2.14 shows HGM distribution from skin to core of the injection molded part in the ND–TD plane. HGMs are seen in this image as black round specs.

A larger magnification is shown in Figure 2.15. In this image, one can see HGMs as round black specs with a lighter core due to the hollow nature of the microspheres. We can also notice empty circular spaces (holes) generated in the microtomed section. These empty holes also represent HGM occupied spaces in the molded part. During 25-μm

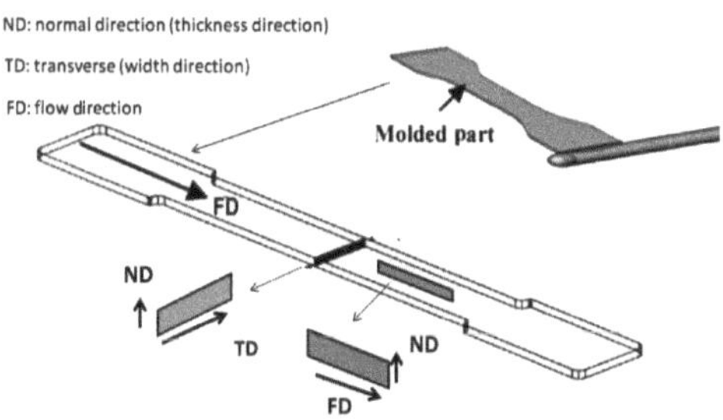

Figure 2.13 Sectioning procedure for optical microscopy.

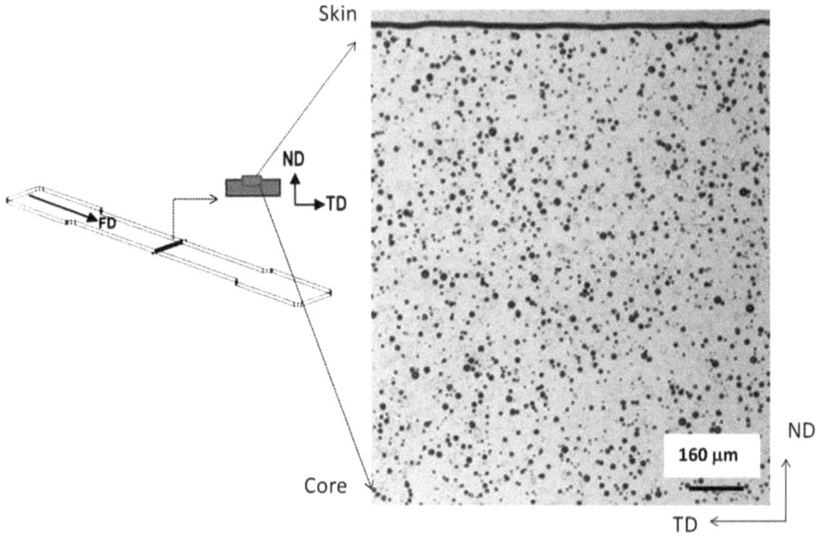

Figure 2.14 Skin to core hollow glass microsphere distribution in injection molded unfilled polypropylene prepared by direct injection molding method.

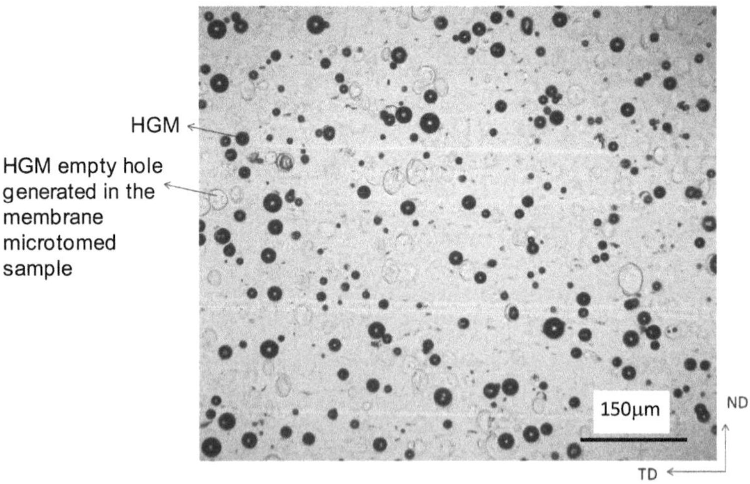

Figure 2.15 A magnified region showing bubble distribution in injection molded unfilled polypropylene prepared by direct injection molding method. HGM, hollow glass microsphere.

sectioning, some HGMs are pulled in with the microtomed section, some are cut along their circles and some stay on the molded part. Therefore, one should note that any measurements made on the individual HGM images do not necessarily reflect the actual diameter or the wall thickness. Only those sections where a HGM has been cut along its equator would provide accurate measurements of wall thickness and diameter. When assessing distribution, one should take into account the empty spaces as HGM occupied spaces as well.

Optical microscopy images are also very useful for distribution of HGMs in the presence of other fillers such as glass fibers in Figure 2.16. Moreover, optical microscopy in polarized form can reveal how the morphological structure of crystalline matrices such as polypropylene is influenced in the presence of HGMs. Figure 2.17 shows HGMs in PP in unpolarized (left) and cross-polarized (right) mode of optical microscopy. One can see in the right cross-polarized image not only the HGMs (black) but also the polypropylene spherulites which are undisturbed by the presence of HGMs.

If higher resolution and three-dimensional (3D) imaging of HGMs are needed using optical microscopy, one of the issues in using conventional optical microscopy on 3D particles, is the shallow depth of field at high magnifications. For example, a 100x objective lens with a numerical aperture of around 1.4 has a depth of field of approximately 1 μm. When

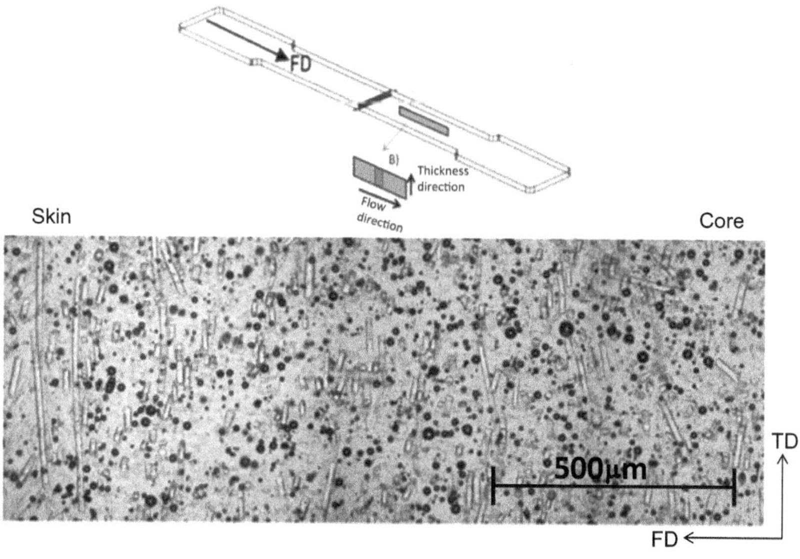

Figure 2.16 Hollow glass microsphere and glass fiber distribution from skin to core in injection molded PP.

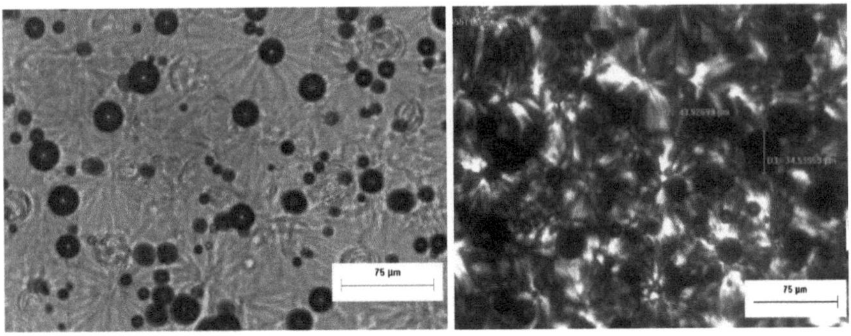

Figure 2.17 Unpolarized (left) and polarized (right) image of hollow glass microspheres in polypropylene.

observing a sample directly, this problem can be circumvented by focusing up and down through the sample. To effectively present a microscopic image of a 3D structure, a technique called focal plane merging or Z-stacking, can be very useful. With this technique, a series of images are captured at different focal depths, and in each image, different areas of the sample will be in focus. The in-focus patches are then blended together (typically using a camera software package) to generate the final image. An interferometry-based technique is applicable only to transparent microspheres such as glass microspheres and some polymer microspheres. A dual-mirror interference microscope is used to image a single HGM that is placed on one of the mirrors. The mirrors are then set so as to be slightly nonparallel which yields an interference fringe pattern of light and dark concentric circles visible on the mirror. The focus is then adjusted until the darkest fringe is positioned on the HGM and the light source is switched from white to monochromatic. By simply counting the number of reference fringes between the darkest fringe outside the HGM and the HGM itself, the wall thickness (t) of the microsphere can be calculated from Eqn (2.17) [7]

$$t = \frac{x\,\lambda}{4(n-1)} \tag{2.17}$$

where

x: the number of fringes;

λ: wavelength of the monochromatic light source; and

N: refractive index of the material (*n*) by using the following equation.

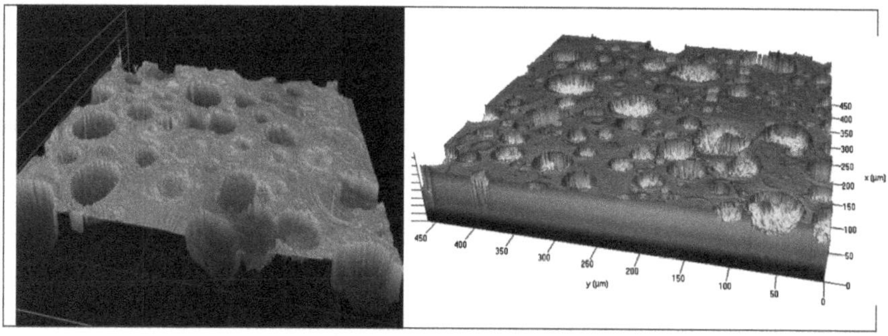

Figure 2.18 Interferometry (left) and laser confocal (right) images of microspheres mounted in epoxy. *With permission from Ref. [7].*

This method has been shown to be accurate to within 0.05 µm for measuring HGM wall thickness [7].

Both laser confocal and interference fringe analysis optical techniques yield similar 3D images of a polished microstructure as shown in Figure 2.18. Both systems yield similar results. Both figures are 3D topologies and could easily be used to measure microsphere diameters and wall thicknesses.

References

[1] B. ASTM, 527: Standard Test Method for Determination of Tap Density of Metallic Powders and Compounds, American Society for Testing and Materials, 2000.

[2] J. Thudium, J. Aerosol Sci. 7 (1976) 167−173.

[3] A. Furuse, December 19, 1989, U.S. Patent 4,888,718.

[4] C.S. Chang, Cereal Chem. 65 (1) (1988) 13−15.

[5] K.B. Carlisle, K.K. Chawla, G.M. Gladysz, M. Koopman, Structure and mechanical properties of micro and macro balloons: an overview of test techniques, J. Mater. Sci. 41 (2006) 3961−3972.

[6] ASTM D3102-78, Practice for Determination of Isostatic Collapse Strength of Hollow Glass Microspheres, 1982 (Withdrawn 1984).

[7] Ji-Z. Liang, Estimation of thermal conductivity for polypropylene/hollow glass bead composites, Compos.: Part B 56 (2014) 431−434.

[8] M. Koopman, G. Gouadec, K. Carlisle, K.K. Chawla, G. Gladysz, Compression testing of hollow microspheres (microballoons) to obtain mechanical properties, Scr. Mater. 50 (2004) 593−596.

[9] T.V. Garza-Cruz, M. Nakagawa, On a hybrid method to characterize the mechanical behavior of thin hollow glass microspheres, Granular Matter 14 (2012) 309–318.
[10] J.Z. Liang, F.H. Li, Simulation of heat transfer in hollow-glass-bead-filled polypropylene composites by finite element method, Polym. Test. 26 (3) (2007) 419–424.
[11] J.Z. Liang, F.H. Li, Heat transfer in polymer composites filled with inorganic hollow micro-spheres: a theoretical model, Polym. Test. 26 (2007) 1025–1030.
[12] J.Z. Liang, F.H. Li, Measurement of thermal conductivity of hollow glass-bead-filled polypropylene composites, Polym. Test. 25 (2006) 527–531.
[13] N. Siu, Leung, et al., Analytical modeling and characterization of heat transfer in thermally conductive polymer composites filled with spherical particulates, Compos.: Part B 45 (2013) 43–49.
[14] H. Zhou, S. Zhang, M. Yang, The effect of heat-transfer passages on the effective thermal conductivity of high filler loading composite materials, Compos. Sci. Technol. 67 (2007) 1035–1040.
[15] J.C. Maxwell, A Treatise on Electricity and Magnetism, Oxford University Press, London, 1892.
[16] H.W. Russell, Principles of heat flow in porous insulation, J. Am. Ceram. Soc. 18 (1935) 1–5.
[17] W.D. Kingery, Conductivity of multicomponent systems, J. Am. Ceram. Soc. 42 (1959) 617–622.
[18] L.E. Nielsen, Thermal conductivity of particulate-filled polymers [J], J. Appl. Polym. Sci. 29 (1973) 3819–3825.
[19] ASTM C518-04 Standard Test Method for Steady-State Thermal Transmission Properties by Means of the Heat Flow Meter Apparatus.
[20] B. Zhu, J. Reinf. Plast. Compos. 31 (19) (2012) 1311–1326.
[21] M.G. Todd, F.G. Shi, Molecular basis of the interphase dielectric properties of microelectronic and optoelectronic packaging materials, IEEE Trans. Compon. Packag. Technol. 26 (3) (2003) 667–672.
[22] Towards model-based engineering of optoelectronic packaging materials: dielectric constant modeling, Microelectron. J. 33 (5–6) (2002) 409–415.
[23] P. Gonon, A. Sylvestre, J. Teysseyre, et al., Dielectric properties of epoxy/silica composites used for microelectronic packaging, and their dependence on post-curing, J. Mater. Sci. - Mater. Electron. 12 (2) (2001) 81–86.
[24] K.C. Yung, Preparation and properties of hollow glass microsphere-filled epoxy-matrix composites, Compos. Sci. Technol. 69 (2009) 260–264.

3 Hollow Glass Microspheres in Thermoplastics

Baris Yalcin and Stephen E. Amos

Introduction

Inorganic solid fillers have greatly contributed to the growth of the thermoplastic industry. Originally fillers were introduced to reduce cost by removing relatively expensive resin. But over time they were recognized as providing functional benefits as well and now are tailored to render plastics with unique properties. Fillers can act as (1) mechanical property modifiers, for example, glass fiber (GF), talc, calcium carbonate; (2) electrical, thermal, and magnetic property modifiers, for example, carbon black, alumina, boron nitride, carbon nanotubes, graphene; (3) surface property modifiers such as silica, molybdenite, graphite, boron nitride; (4) fire retardants, for example, metal hydroxides; and (5) processing aids and stabilizers such as fumed silica and hydrotalcites. In most cases, fillers modify more than one property.

Solid fillers have a density higher than that of the host resin and add significant amount of weight to the final plastic part. This is especially true for highly filled situations such as the flame retardant and thermally conductive applications referenced above. For example, aluminum trihydrate is used as a flame retardant and the loading can exceed 70 wt%. The advent of hollow glass microspheres (HGMs) in the 1960s changed the paradigm that fillers cause the weight of the plastic composite to increase. This also required a different approach for formulating—namely volume formulating, which will be demonstrated in upcoming examples.

Reducing the weight of thermoplastics parts has been a high-priority objective in various industries such as transportation, aerospace, handheld electronics, and sports and leisure. Automotive plastics have been extensively used for years to replace metal parts and cut weight to improve Corporate Average Fuel Economy (CAFÉ) levels, compared to those of a generation ago. According to the U.S. Department of Energy, "...for every 10% of weight eliminated from a vehicle's total weight, fuel economy improves by 7%" [1]. HGMs are currently used in a variety of lightweight automotive applications, including thermoplastics, sheet and

Hollow Glass Microspheres for Plastics, Elastomers, and Adhesives Compounds
http://dx.doi.org/10.1016/B978-1-4557-7443-2.00003-7

bulk molding composites, underbody coatings (plastisols), structural foams, and auto body fillers. HGMs are excellent strength/weight optimizers when they are used in filled polymer systems such as GF, talc, and calcium carbonate filled thermoplastics. Reducing and replacing a certain percentage of these high-density fillers with HGMs results in weight reduction while significantly maintaining the original mechanical properties of the composite. For instance, thermoplastic olefin (TPO)-based compositions containing large amounts of talc have been successfully modified with HGMs, reducing the density up to 13% while maintaining an acceptable balance of performance and processing characteristics for injection molded automotive parts [2]. HGMs have also been shown to be successfully incorporated into high-temperature polymers, such as polyetherimide for 10+% weight reduction and greater savings in aerospace applications [3]. In comparison with GF reinforced grades, significant weight savings were achieved with price advantages.

Benefits of HGMs in Thermoplastics

HGMs impart several benefits to thermoplastics in addition to density reduction. These include:

- productivity benefits through faster cooling rates from the melt
- dimensional stability (sink and warpage elimination)
- increased stiffness (modulus) and heat distortion resistance
- reduced thermal conductivity and dielectric constant

All of these new functions and benefits can be achieved with class-a surface and with existing equipment enabling new design functions.

Productivity Benefits of HGMs Through Faster Cooling Rates from the Melt

Cooling of the thermoplastic parts from the melt is a very important factor in the economics and operation of the process. Long cooling times incur additional manufacturing costs and can limit production capacity. The cooling time of molten plastic in the process can be estimated by calculating thermal diffusivity (α), shown in Eqn (3.1). This material property is a measure of a material's ability to transmit heat relative to its ability to store heat [4]. All process parameters kept constant, materials with higher thermal diffusivity require shorter cooling times.

$$\propto \;=\; \frac{k}{\rho\, c_p} \tag{3.1}$$

where

$\propto$ = thermal diffusivity

k = thermal conductivity

ρ = density

c_p = Specific heat

HGMs increase cooling rates of thermoplastic parts from the melt through their effect on thermal diffusivity. Increased HGM weight fraction decreases composite density and composite specific heat capacity which in turn increases thermal diffusivity and hence cooling rates. Composite density decreases because HGMs are lower in density than polymers. Composite specific heat capacity decreases because the specific heat capacity of glass [5] (~750 J/kg K) is lower than that of most thermoplastic materials (1500–3500 J/kg K) [6]. HGMs also influence thermal conductivity which has a proportional effect on thermal diffusivity. Fillers with high thermal conductivity increase composite thermal conductivity and hence the thermal diffusivity of the parts. HGMs of very low densities (0.12–0.38) could decrease thermal conductivity of polypropylene (PP) which has a thermal conductivity of about 0.21 W/m K. However, in injection molding processes, HGMs with densities between 0.46 g/cc and 0.6 g/cc are typically used due to the strength requirements. HGMs, at this density range, have neutral to minimal effect on thermal conductivity of polymers such as PP and polyamides (PA).

Case studies were reported that calculated the cooling rate effect of HGMs in injection molded or extruded parts through thermal imaging [7]. An experimental setup with an IR camera was constructed to take thermal images of injection molded PP and predict the cooling time from Eqn (3.2), which is typically used for plate type injection molded geometries [8].

$$t_c \;=\; \frac{h^2}{\pi^2 \alpha} \ln\left(\frac{4}{\pi}\,\frac{T_{\text{melt}} - T_{\text{mold}}}{T_{\text{eject}} - T_{\text{mold}}}\right) \tag{3.2}$$

where

t_c: theoretical minimum cooling time

h: thickness of the part

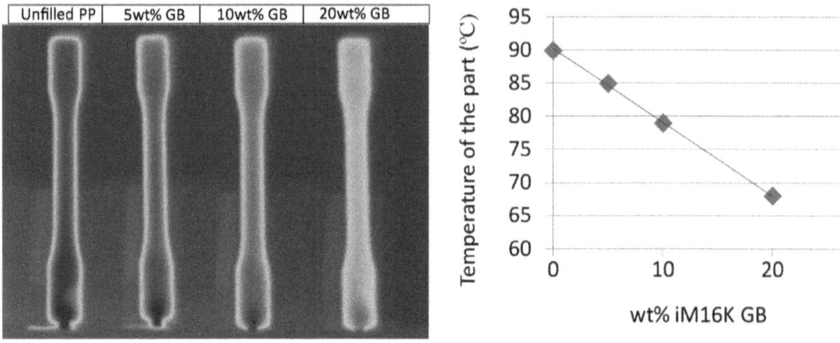

Figure 3.1 Area average temperatures of the polypropylene (PP) molded parts after ejection as a function of hollow glass microsphere loading (with permission from ref. [7]).

α: thermal diffusivity of the part

T: temperature

As shown in Figure 3.1, the temperature of the ejected part was reduced from 90 °C to 68 °C when the HGM (3M™ Glass Bubble iM16K, 0.46 g/cc) loading was increased from 0 wt% to 20 wt%; evidence that the parts cool faster with HGMs. Figure 3.1 also shows that there is quite a linear relationship between the weight percent HGM loading and the temperature of the ejected part. Reduction in temperature is roughly 1.1 °C per each wt% of HGM (at a density of 0.46 g/cc) added into the formulation.

In the same study, cooling time studies were compared in filled systems. For talc-filled PP, a 20 wt% talc-containing formulation was compared to a formula containing 10 wt% talc and 4 wt% HGMs. For GF filled PP, a 15 wt% GF-containing formulation was compared to one formulation with 15 wt% GF and 5 wt% HGMs, and another with 18 wt% GF and 7 wt% HGMs. These formulations were chosen as they demonstrate comparable mechanical properties to the original formula with reduced densities, a key benefit of HGMs. The results are summarized in Table 3.1.

3M reported another mold cycle time analysis for PA6 composites through an independent study by SKZ Institute in Germany by measuring the ejection temperature of a 60 × 60 × 2 mm molded part and lowering the cooling time to match the same temperature measured without the HGMs. The experiment setup is shown in Figure 3.2.

Table 3.1 Calculated Cooling Times to Reach a Temperature of 90 °C (Unfilled Systems), 88 °C (Glass Fiber (GF) Filled), and 83.3 °C (Talc Filled) in the Presence of Hollow Glass Microspheres (ref. [7])

Material	Hollow Glass Microsphere Loading Wt%	Other Fillers Wt%	Cooling Time (s)	Cooling time Reduction (%)
Unfilled polypropylene	0	–	16	–
	5	–	14.7	8.2
	10	–	13.1	18
	20	–	10	37
GF filled PP	0	15 GF	16	–
	5	15 GF	13.5	15.3
	7	18 GF	12.15	24
Talc-filled PP	0	20 Talc	16	–
	4	10 Talc	15	6

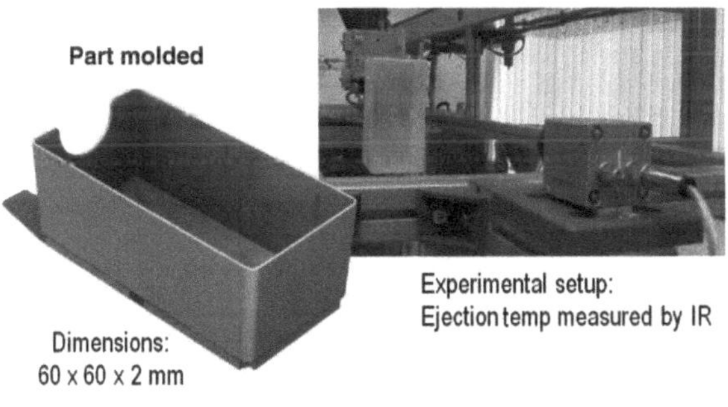

Figure 3.2 Experimental setup and part dimensions used in mold cycle time analysis for polyamide (with permission from ref. [7]).

In the case of PA6, mold cycle time was reduced 12% when 7 wt% (16 vol%) HGM (0.46 g/cc) was added. At 15 wt% GF and 4 wt% HGM (0.46 g/cc), the reduction in cycle time was 5% compared to unfilled PA6.

Reduced cooling in the presence of HGMs was also shown for extruded profiles of polymer wood composites (PWC) [9]. HGMs in this case remarkably increased the rate of cooling of the extruded profiles as shown in Figure 3.3, by reducing the total thermal mass. One may anticipate that the improved rate of cooling would be very useful in profile extrusion. Recall that the relatively low rate of cooling of the extruded profiles has been one of the major bottlenecks in profile extrusion. The results suggested that introducing HGMs might significantly improve the rate of profile extrusion. This can significantly increase machine and operator productivity providing cost savings not initially factored in to the cost of the compound.

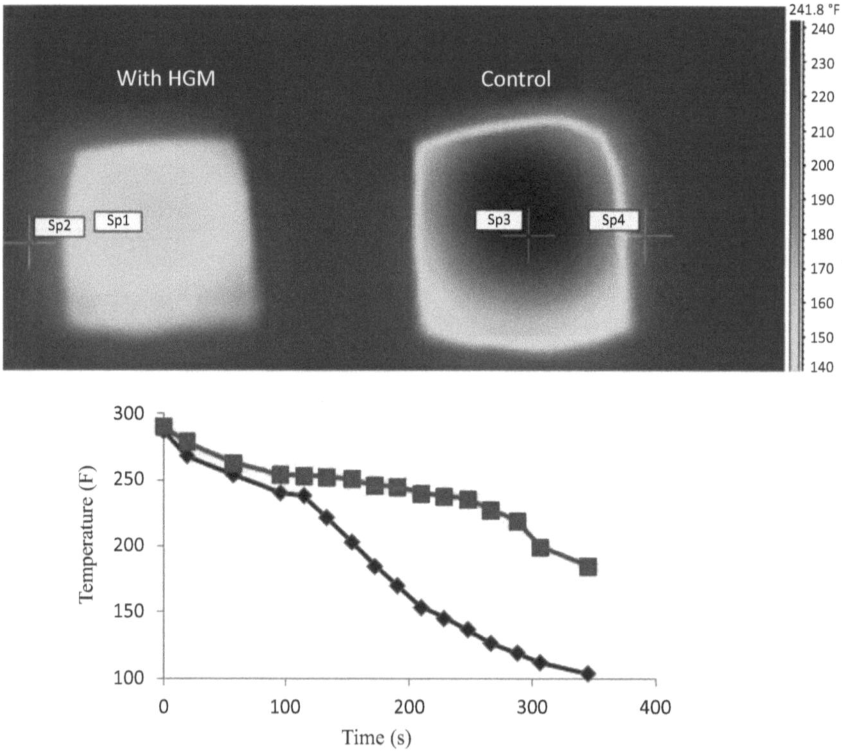

Figure 3.3 The measured temperatures of the unfilled (red) and 10 wt% hollow glass microsphere filled (blue) extruded profile sheets upon cooling from melt. Top: Representative IR camera images (with permission from ref. [9]).

Dimensional Stability

In polymer processing operations, dimensional stability issues such as sink, shrinkage, and warpage constitute a major problem resulting in decreased productivity. There are several factors that affect dimensional stability including part design, material and mold design, and process conditions [10]. HGMs have an effect on the material design through their spherical 1:1 aspect ratio geometry. HGMs, especially at higher loadings, allow the material to cool homogeneously preventing occurrence of sink marks typically observed in thick parts or features such as ribs and bosses that cool more slowly than neighboring regions.

Mold shrinkage (in-mold shrinkage or molded part shrinkage), although a volume phenomenon, usually refers to the difference between the linear dimension of the mold at room temperature and that of the molded part at room temperature within 48 h following ejection. All polymers that cool from the melt will shrink due to density changes but shrinkage is more apparent in crystalline polymers than in amorphous polymers. A common misunderstanding is that the shrinkage values are a direct indication of potential part warpage. Warpage, a distortion of the shape of the final injection-molded item, is caused by differential shrinkage; that is, if one area or direction of the article undergoes a different degree of shrinkage than another area or direction, the part will warp. Postmold shrinkage is another common shrinkage term. It refers to any additional shrinkage that occurs after the initial 48 h period.

Fillers influence the shrinkage by offsetting some volume of polymer with a low-shrinking filler particle. The shrinkage of resins containing isotropic fillers, such as HGMs, will be more isotropic than resins containing high aspect ratio fillers like fibers or platelets. This results from orientation of the fillers in the flow direction during filling, and the restricted shrink along the long axis of the filler particles. Fibers are known to create excessive warp as the restricted shrink in the flow direction is compensated by an increased shrink of the polymer in the transverse direction. Anisotropic shrinkage of fiber-reinforced polymers can be attributed to the fact that the fibers become oriented in the flow-shear field during injection molding. Unlike polymer molecules that can orient and relax during filling and cooling, fibers have no tendency to reorient in the cooling melt. Flow-induced fiber orientation is maintained during polymer cooling. Both shear and elongational flow will influence the orientation of fiber reinforcements. Processing variables such as fill rate, cavity thickness, melt viscosity, and gating scheme are all significant factors affecting fiber orientation.

Table 3.2 Cycle Time Analysis for Polyamide

Material	Total Cycle Time t_G [s]	Cycle Time Reduction [%]
PA6	40.2	–
PA6 HGM–16 v%	35.2	12
PA6 GF 15–6 v% HGM–10 v%	38.2	5

PA, polyamide; GF, glass fiber.

Table 3.3 shows in-flow and cross-flow shrinkage and Figure 3.4 shows a graph of differential shrinkage for injection molded PA6 with HGMs and GFs [11]. Table 3.4 and Figure 3.5 show the same for Talc and HGM filled copolymer PP. One can see that the differential shrinkage (warpage) is reduced in HGM filled formulations, the most noticeable decrease being in the GF filled PA6. Differential shrinkage in talc filled PP is also reduced. In both the PP and PA cases, one can also notice that compared to GF or talc only formulas, the shrinkage with HGMs is closer to that of the unfilled resins especially at low loadings as shown in these examples. In other words, the level of shrinkage is less with HGMs but more uniform when differential shrinkage is considered. This is an advantage when considering that no major tool design changes are necessary to adjust shrinkage in unfilled PP or PA applications when low levels of HGMs are used.

Processing of HGMs

In this section, we will review how HGMs can be incorporated into melt-processable thermoplastics.

Similar to most other fillers, HGMS usually need to be compounded into a polymer before being used in a post processing method such as injection molding, film blowing, and so on. Direct use of HGMs, that is, without compounding, in these post processing polymer operations requires specialized techniques and equipment which we will also briefly touch upon.

Most frequently, HGMs are precompounded into a final formula in a twin screw extruder (TSE) or compounder before being used in a post processing operation to form an article (Figure 3.6 Top). Often, they can also made into a concentrated masterbatch form (at ~50 vol%) in

Table 3.3 In-flow and Cross-flow Shrinkage in Injection Molded Polyamide 6 (PA6) with Hollow Glass Microspheres (HGMs) and Glass Fibers (GFs) (Schulamid NV12)

Material	Hold Pressure p_N [bar]	Processing Shrinkage [%]		Postmolding Shrinkage		Total Shrinkage	
		In-flow	Cross-flow	In-flow	Cross-flow	In-flow	Cross-flow
PA6	150	1.42	1.72	0.002	−0.007	1.43	1.72
PA6	250	1.39	1.55	0.001	−0.007	1.39	1.54
PA6 + HGM 16 vol% (7 wt% HGM)	150	1.53	1.57	0.006	−0.005	1.54	1.56
PA6 + HGM 16 vol% (7 wt% HGM)	250	1.54	1.36	0.003	−0.005	1.54	1.35
PA6 + GF 6 vol% + HGM10 vol% (15 wt% GF, 4 wt% HGM)	150	0.93	1.21	0.010	−0.005	0.94	1.20
PA6 + GF 6 vol% + HGM10 vol% (15 wt% GF, 4 wt% HGM)	250	0.87	1.08	0.012	−0.006	0.88	1.08
PA6 + GF16 vol% (30 wt% GF)	150	0.41	0.96	−0.014	−0.007	0.40	0.95
PA6 + GF16 vol% (30 wt% GF)	250	0.41	0.87	0.009	−0.008	0.42	0.87

Courtesy of 3M

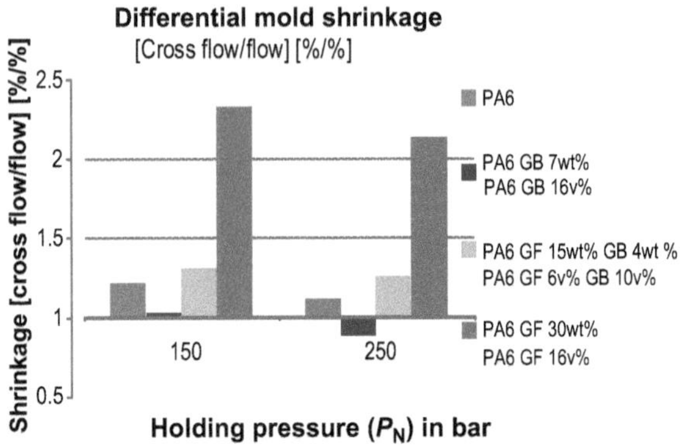

Figure 3.4 Differential Mold Shrinkage in injection molded PA6 with hollow glass microspheres and Glass Fibers (PA6 Schulamid NV12) (with permission from ref. [11]).

a certain carrier polymer resin in a TSE and then let down at the post processing step by mixing with the polymer resin in a post processing step (Figure 3.6).

Whether HGMs are precompounded into a final formula or in a masterbatch form first, twin screw extrusion is an important and critical first step where maximum survival rate of HGMs is desired.

Corotating intermeshing TSEs are typically used for compounding HGMs into polymers in a continuous manner. HGMs are preferably introduced downstream in the extruder via side stuffing or top feeding ports into a fully molten polymer stream [12]. This is similar to GF feeding where the fiber attrition is kept to a minimum by downstream addition [13]. Figure 3.7 shows a TSE configuration suitable for compounding HGMs either alone or in the presence of other fillers.

In the configuration shown in Figure 3.7, polymer resin is starve-fed in zone 1 via a resin feeder and passed through a set of kneading blocks to ensure its complete melting before HGMs are introduced downstream (zone 4). HGMs should be starve-fed into a side feeder via a supply feeder. It is crucial that conveying elements with high free volumes generated by deeply cut screw channels (outer diameter/inner diameter (Do/Di) ratio: 1.7 to 1.9 or more) be used. As Do/Di increases, channel depth (h) increases. Increased channel depth translates into higher free volume necessary to accommodate HGMs and lower shear rates as shown in Figure 3.8 [14]. This is especially important for high loadings of low-density HGMs at collapse strengths of 5000 psi to 8000 psi. As the HGM collapse strength increases, the above suggestions become much less stringent.

Table 3.4 In-flow and Cross-flow Shrinkage in Injection Molded Polypropylene (PP) Copolymer (PP C080MT) with Hollow Glass Microspheres (HGMs) and Glass Fibers

Material	p_N [bar]	Processing Shrinkage [%]		Postmolding Shrinkage [%]		Total Shrinkage [%]	
		In-flow	Cross-flow	In-flow	Cross-flow	In-flow	Cross-flow
PP copolymer	100	1.68	1.73	0.020	0.016	1.70	1.75
PP copolymer	200	1.41	1.47	0.028	0.020	1.44	1.49
PP copolymer +4 wt% HGM im17K	100	1.61	1.67	0.018	0.011	1.62	1.68
PP copolymer +4 wt% HGM im17K	200	1.37	1.41	0.022	0.003	1.40	1.42
PP copolymer +10 wt% talc +3, 7 wt% HGM im17K	100	1.42	1.55	0.025	0.010	1.44	1.56
PP copolymer +10 wt% talc +3,7 wt% HGM im17K	200	1.23	1.27	0.021	0.008	1.25	1.28
PP copolymer +20 wt% talc	100	1.18	1.33	0.029	0.005	1.21	1.34
PP copolymer +20 wt% talc	200	1.04	1.09	0.019	0.012	1.06	1.10

Courtesy of 3M

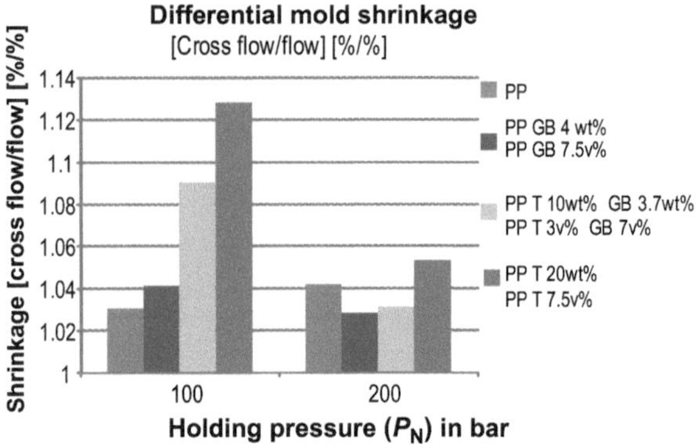

Figure 3.5 Differential mold shrinkage in injection molded PA6 with hollow glass microspheres and glass fibers (with permission from ref. [11]).

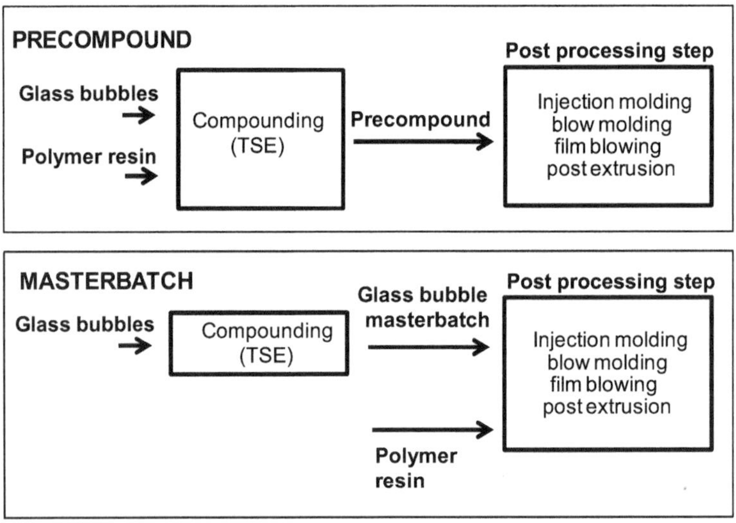

Figure 3.6 Incorporating hollow glass microspheres into polymer articles injection molded, blown film, or extruded profiles. Top: Precompound approach. Bottom: Masterbatch approach.

One of the advantages of HGMs during compounding is their ability to distribute in the molten polymer without having to resort to aggressive kneading and distributive mixing elements. In fact, simply through mere friction from the barrel wall and conveying elements, HGMs distribute reasonably well. However, further downstream in the process, a short set

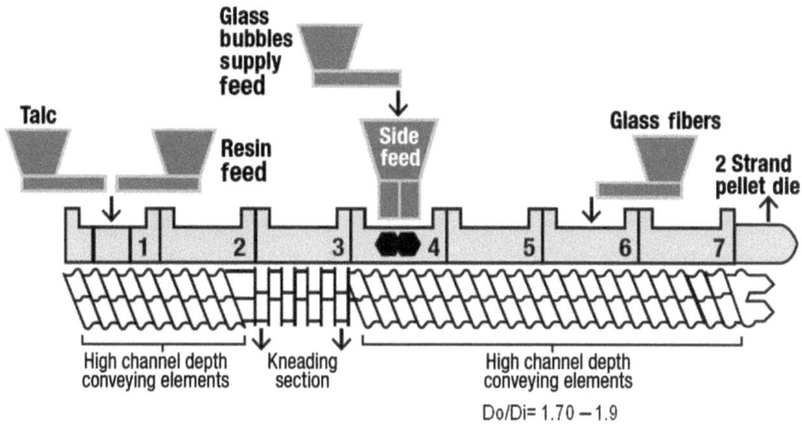

Figure 3.7 Corotating twin screw extrusion configuration for compounding hollow glass microspheres alone or in the presence of other fillers.

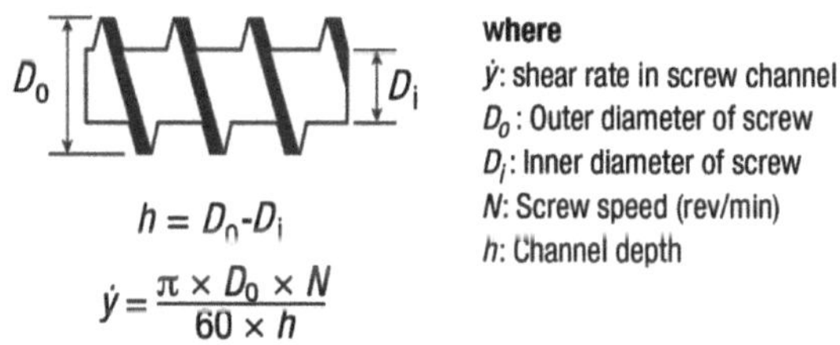

Figure 3.8 Effect of channel depth and screw speed on shear rate.

of distributive elements can be used if necessary, especially at very low loadings of HGMs. At high loadings, such as 50 vol%, HGMs occupy the entire resin at their closest packing configuration making distribution irrelevant. Venting, following a reverse element, is optional depending on the application before the compounded material is discharged. If the compounded pellets are to be subsequently injection molded, venting is not crucial since the small amounts of air trapped during compounding can escape through the vents during injection molding.

HGMs can also be compounded along with various fillers such as GF, talc, or clay as shown in Figure 3.7. Platy fillers (e.g., clay, talc) that need dispersion are typically added before the kneading block to facilitate their dispersion. Talc can be added either in zone 1 or into the side feeder with the HGMs through a different supply feeder. Dry blending talc with HGM powder first and then side feeding the mixture is also possible if there are not enough supply feeders. Since both HGM and talc are in powder form, they do not phase separate in spite of the differences in density. However, this should be checked for each supply and side feeder auger system used. GFs cannot be dry blended with HGM powders since they phase separate quickly due to the differences in their physical forms, that is, chopped fibers versus powder. GFs can be added downstream or upstream of the HGMs. Mica, due to its higher aspect ratio, needs to be side-fed similar to HGMs.

It is important to determine HGM survival after compounding. We have discussed the methodologies to determine HGM survival in Chapter 2 and therefore are not reviewing it here.

Configuring the appropriate compounding system is the first step to achieving high HGM survival. The next step is to understand the process and polymer resin material parameters.

Pelletizing Effect on HGM Survival

Most HGM compounded polymers need to be pelletized for further processing, for example, for injection molding. In a standard water bath pelletizing system, the strands after cooling enter into the cutting chamber of the pelletizer where a rotating blade cuts the strands into small pellets. In a standard pelletizer, a certain amount of HGM breakage is possible depending on the HGM collapse strength (Figure 3.9).

The small amount of HGM breakage that is observed with low-strength HGM grades during standard pelletization can be minimized, if not prevented, by an underwater pelletizer. In this process, the molten polymer is cut into droplets by the fast rotating blades of the pelletizer just as it is

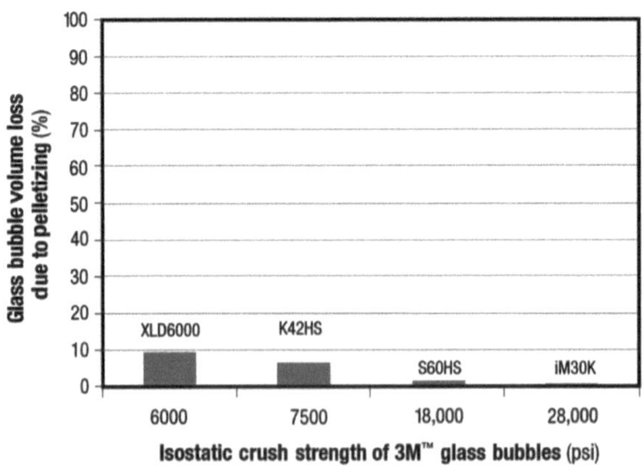

Figure 3.9 Percent of hollow glass microspheres (HGM) void volume loss due to pelletizing as a function of isostatic crush strength of HGM used in homopolymer polypropylene with a melt flow index of 4 g/10 min at 230 °C (with permission from ref. [12]).

exiting the die hole and emerging into the process water. Since the polymer is cut when the polymer is molten, the bubble breakage is prevented.

Effect of Polymer Melt Viscosity on HGM Survival

It is well known from studies on GF attrition during compounding that increased polymer melt viscosity causes GF breakage. HGMs show the same trend when lower strength HGM grades (6000 psi collapse strength) are used under aggressive compounding conditions. Higher melt viscosities result in higher shear and compressive stresses which increase the possibility for HGM breakage. Figure 3.10 shows % HGM void volume loss in PP as a function of melt flow index (MFI).

With a high MFI (low viscosity) PP polymer, HGM breakage was negligible (3.55 vol%). The breakage rate was much higher (25.6%) when a lower MFI (high viscosity) PP was used. The effect of melt viscosity on HGM breakage becomes less important when high-strength HGM grades are used (10,000 psi and higher) but it is still recommended to use a lower viscosity polymer if a choice can be made.

Effect of Back Pressure on HGM Survival

Back pressure is one of the most critical parameters that influence HGM survival. In extrusion, back pressure is the amount of resistance applied to the melt which can be caused by the presence of downstream

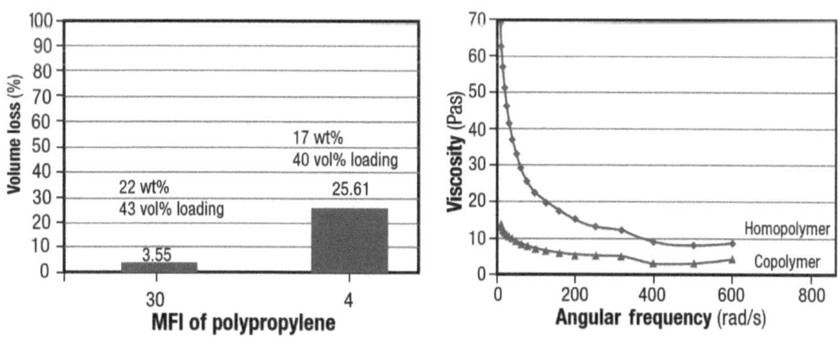

Figure 3.10 HGM (6000 psi collapse strength) % volume loss in polypropylene as a function of melt flow index (MFI). LyondellBasell Pro-fax™ 6523. MFI (230 °C/2.16 kg): 4 g/10 min and LyondellBasell Pro-fax™ SG899 MFI (230 °C/2.16 kg): 30 g/10 min. Shear rate dependent viscosity data of neat resins are shown in right graph (with permission from ref. [12]).

equipment such as screens, dies, and so on. In injection molding, it is the resistance applied to the rear of the screw as it rotates and collects the melt in front of the screw. In either case, at constant screw speed, increasing back pressure compresses the melt increasing friction and shear applied to the material. Increased friction and shear can lead to HGM breakage.

Figure 3.11 shows void volume loss due to HGM breakage and final density of the HGM (0.318 g/cc-isostatic crush strength of original density 6000 psi) during compounding with 6523 MFR-4 PP with (1) no die, (2) die with a three hole strand, and (3) die with a two hole strand. When a strand die with two holes is employed, 14% breakage is calculated in a high-viscosity PP at 15 wt% (30 vol%) HGM loading. By simply opening another hole in the strand die, percent HGM void volume loss drops to 9.8%. When the die is removed and the extrudate is simply collected at the large opening, the HGM breakage further reduces to 5.5% which results in a final density of 0.332 g/cc for the HGMs as determined from ash analysis described in Chapter 2. This example shows the effect of

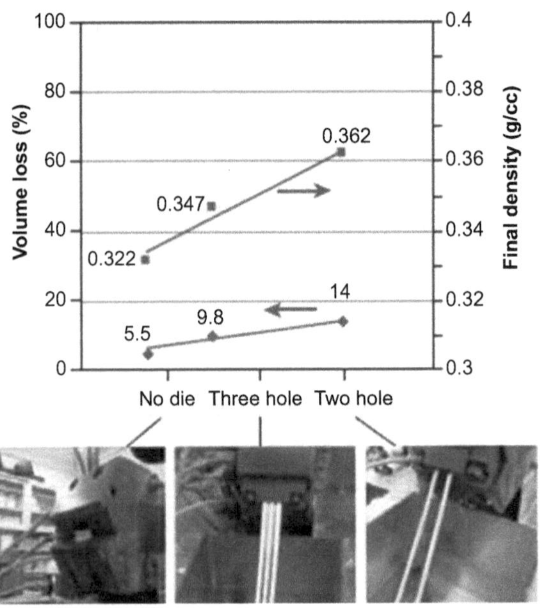

Figure 3.11 Percent void volume loss due to HGM breakage and final density of 15 wt% (~30 vol%) hollow glass microsphere (3M™ Glass Bubble XLD6000 0.318 g/cc, 6000 psi) in 6523 (4 melt flow rate homopolymer PP) with no die, die with a three hole strand die, and with a two hole strand die (with permission from ref. [12]).

back pressure on the survival of HGMs and importance of die design. Increasing the number of holes and/or increasing their diameter decreases back pressure and helps minimize bubble breakage. However, for a constant volumetric flow rate, it also slows down the flow of polymer coming out of the die, that is, strand output velocity slows down. When the velocity is too slow, it becomes difficult to synchronize pelletizing with the slow strand speed. Therefore, one must optimize die design while keeping melt handling issues in mind. Similarly, one can imagine the effect of screens with different mesh sizes. Larger openings in the screens result in lower back pressure minimizing HGM breakage.

Effect of HGM Loading on HGM Survival

Another important factor is the amount of HGM loading in the extruder. As mentioned above, high-channel depth screw elements are necessary to accommodate large loadings of HGMs. For a constant channel depth, high loading of HGMs increases the probability of microsphere to microsphere contact and hence breakage. This is true in the case of low density low strength HGMs but not affected when high-strength HGMs are used as shown in Figure 3.12. In order to minimize breakage for high loadings of low strength HGMs, it is recommended that HGMs be added in more than one zone downstream.

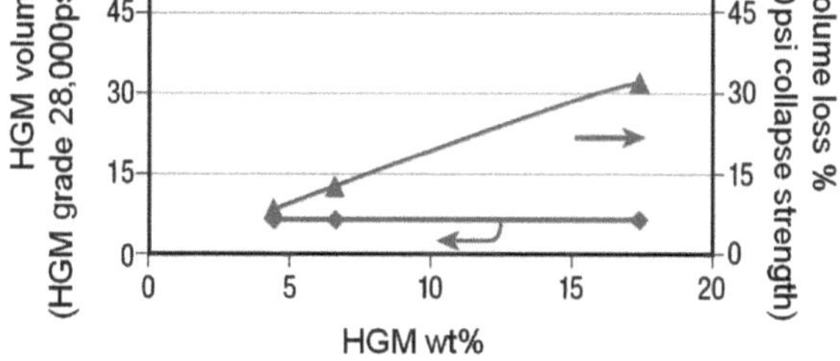

Figure 3.12 Percent hollow glass microsphere (HGM) void volume loss as a function of loading in a 6523 polypropylene (PP) homopolymer with melt flow index (MFI) 4 g/10 min (with permission from ref. [12]).

Summary of important points to consider during compounding with HGMs:

1. Twin screw corotating intermeshing extruders are recommended for compounding HGMs.
2. It is highly recommended that HGMs be added into an already molten polymer at a downstream port via a side or top feeder (side feeder is preferred).
3. A side feeder should be fed via a supply feeder. This will ensure starve feeding of bubbles into the polymer melt and allow various volume % loadings to be prepared. If the bubbles are flood-fed into the hopper of a side feeder, clogging and bridging may occur.
4. The extruder should have a high free processing volume accomplished by deeply cut screw channels with an outer diameter/inner diameter (Do/Di) ratio of 1.70 or more.
5. Preheating of HGMs, although not mandatory, could help prevent rapid temperature decrease of the polymer melt, which could cause rapid increase in viscosity. It is advised that the HGMs be preheated or higher temperatures are used at the side feeding ports.
6. After the HGMs are added into the molten polymer, they should be conveyed via standard conveying screw elements for a while before entering distributive block sections (if any need to be used).
7. Inlet design of the side feeder into the extruder is very important, especially if high volume percentages of HGMs are formulated. The screw elements in the inlet section should be of the conveying type with a very high OD/ID ratio, such as 1.70 or more.
8. Minimal back pressure is preferred during compounding with HGMs. In this respect, a die design that creates low back pressure is important. Likewise, screens with too large mesh sizes should be avoided.
9. An underwater pelletizer is the preferred method of pelletizing and should be used especially when compounding low density, low strength HGMs.
10. If possible, resin parameters should be considered to prevent breakage—lower viscosity, higher MFI resins are preferred as well as materials that are softer and more elastic.

Most fillers (GF, talc, mica, and so on) used are greater than 2.5 g/cc, and formulating with weight percentages is very common. However, as will be apparent in the following sections, it is important to understand formulating using volume percentage when HGMs are involved. For instance, if a formulator decides to replace 30 wt% GFs (2.54 g/cc) in PA66 with 30 wt% HGMs (0.6 g/cc), they would be making two big mistakes. First, they would be using 85 vol% polymer resins in the GF case and only 55 vol% in the HGM case. In addition, due to the differences in morphologies, mechanical properties would be significantly different. The following sections are intended to shed some light on effective formulating with HGMs in two thermoplastic resin systems that are popular in transportation applications—polyolefins and nylon. When done well, the physical properties of the HGM containing formulations are similar to the originally selected material without HGMs. Examples in other polymers are also provided.

HGMs in Polyolefins

Polyolefins are the largest group of thermoplastics and a very important class of commercial polymers being used in a wide range of applications. Polyolefins are polymers of alkenes with the general formula C_nH_{2n}, such as ethylene, propylene, butenes, isoprenes, and pentenes, and copolymers and modifications thereof. The two most important and common types of polyolefins are polyethylene (PE) and PP. An inherent characteristic common to all polyolefins is a nonpolar, nonporous, low-energy surface that is not receptive to inks, and lacquers without special oxidative pretreatment. Polyolefins are processed by various conversion processes including extrusion, injection molding, blow molding, and rotational molding methods. Thermoforming, calendering, and compression molding are used to a lesser degree.

Starting in 1950, new catalysts for olefin polymerization were discovered and the development led to both an improvement in the quality of polyolefins and also to diversification of their applications [15]. The individual members of the polyolefin family offer a fairly broad spectrum of structures, properties, and applications. This spectrum can be broadened even further by blending polyolefins of different types (e.g., PE/PP). Furthermore, many other polymers can be improved by adding polyolefins to them and by compatibilization (e.g., PP/Polystyrene/Styrene-Ethylene-Propylene diblock).

TPOs refer to three-phase polyolefin/rubber/filler blends commonly used by the plastics manufacturers and processing tiers. In some TPO formulations, rubber and/or filler can be omitted depending on the end application. The polymer phase is typically based on PP, copolymer PP, or in some occasions PE. These polymers are chosen as the matrix phase due to their low cost, ease of processability, and wide range of properties that can be adapted by the resin chemistry and/or additives.

Common rubbers in a TPO formulation include EPR (Ethylene propylene rubber), EPDM (EP-diene rubber), EO (ethylene–octene), EB (ethylene–butadiene), SEBS (Styrene–ethylene–butadiene–styrene). Rubbers in a TPO formulation improve impact properties of the PP phase which is typically low, especially at low temperatures.

Common fillers in a TPO formulation include, though are not restricted to talc, mica, GF, carbon fiber, wollastonite, and metal oxy sulfate. Fillers are mainly used to stiffen and reinforce the TPO blend, that is, increase tensile and flexural strength and modulus. Fillers also increase the heat distortion temperature (HDT).

Table 3.5 summarizes the four main components of TPOs and their functions.

HGMs require careful formulation in TPOs to maintain an acceptable balance of weight, performance, and processing characteristics for molded

Table 3.5 Thermoplastic Olefin Components and Functions

Component	Example	Function
Polymer	PP (Polypropylene)	Main matrix
	PE (polyethylene)	
Elastomer	EPR (Ethylene propylene rubber)	Improve cold temperature impact properties
	EPDM (EP-diene rubber)	
	EO (ethylene–octene),	
	EB (ethylene–butadiene)	
	SEBS (styrene–ethylene–butadiene–styrene)	
Reinforcing filler	Talc, Nano clay, mica	Increase stiffness (strength, modulus), heat distortion temperature
	Glass fiber (short, long), wollastonite, whiskers, ceramic fibers	
Additives	Pigments, stabilizers	UV, heat, etc.

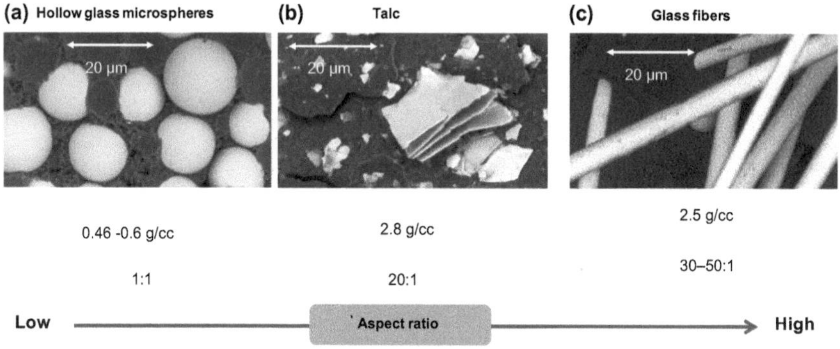

Figure 3.13 Scanning electron microscopy of (a) hollow glass microspheres, (b) talc and (c) glass fibers (with permission from ref. [11]).

parts. For an effective formulation, it is important to understand how HGMs influence properties compared to typical high aspect ratio fillers used in TPOs or other systems. The core difference comes from the geometrical shape. Figure 3.13 shows HGM shape and aspect ratio compared to various other fillers. HGMs are geometrically isotropic fillers with an aspect ratio of one and have intrinsically low surface areas due to sphericity. This is the main reason why they can be incorporated at very high volume loadings compared to inorganic particles with high geometrical anisotropy (e.g., GF, talc, and so on) and still exhibit acceptable viscosity for further polymer processing and shaping operations. For instance, PP with 45 vol% talc (corresponds to 72 wt%) would be practically impossible to process due to viscosity increase, whereas HGMs at the same volume loading can still flow and be processed.

On the other hand, fillers with high geometrical anisotropy are more efficient for reinforcing composites compared to HGMs, that is, increase modulus and strength more efficiently than HGMs. This is because fibrous (e.g., GF, wollastonite) and platy (e.g., talc, mica, nanoclay) fillers have large aspect ratios (20–50) and surface areas and tend to align preferentially in the process flow direction. These fillers also cause orientation of the polymer molecules during high shear processes such as injection molding. The stiffening action of HGMs, on the other hand, comes from the resin space that they occupy and not from oriented structures that they impart to the polymer resins. Preferential orientation seen in high aspect ratio fillers can introduce challenges in dimensional stability such as increased shrinkage in in-flow rather than cross-flow direction causing warpage. HGMs, when used along with other reinforcing fillers, can

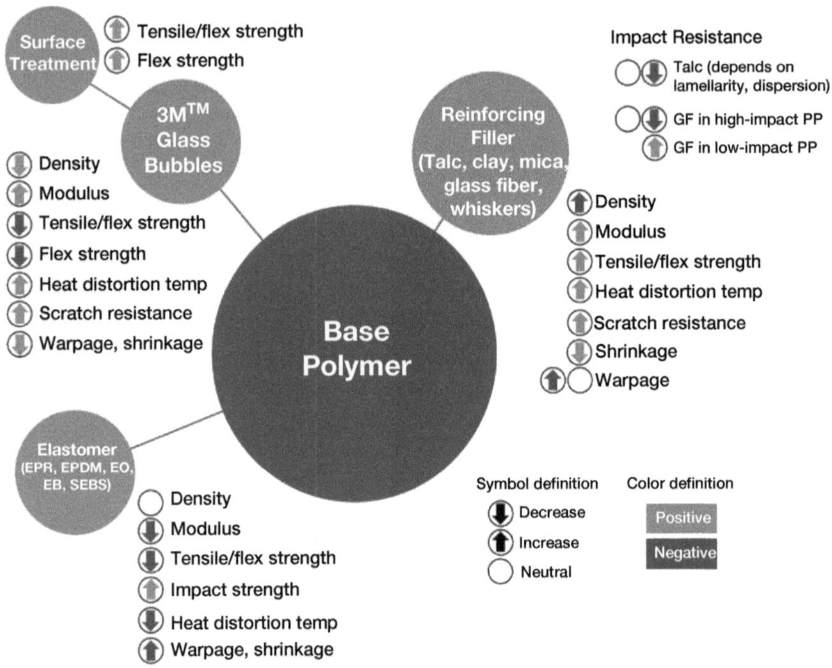

Figure 3.14 Effect of major thermoplastic olefin components on final composite properties (with permission from 3M).

provide excellent dimensional stability characteristics due to their isotropic nature. Because HGMs and reinforcing fillers impart different attributes to a polymer, it is typically not recommended to make a one to one volume replacement of reinforcing fillers with HGMs.

Figure 3.14 shows the effect of major components on TPO properties. Green indicates a positive attribute while the red color indicates a negative influence on properties. Density increase, in this case, is regarded as a negative attribute as lightweighting is typically considered valuable for sustainability. It is important to emphasize that Figure 3.14 only shows whether a property will increase or decrease but does not indicate to what extent the increase or decrease in property will take place. The base polyolefin (in this case PP) that forms the framework of the TPO has a large influence on final TPO properties. Elastomers are mainly used for cold impact strength and are a crucial part of the TPO properties but they do have a negative effect on most other properties. If impact requirements of the specification can be achieved by the base PP, elastomers can be removed from the formula. Reinforcing fillers have several positive property attributes for a TPO compound including increases in stiffness (modulus and tensile/flex strength), heat distortion, scratch resistance (especially with GFs), and decreased shrinkage. In certain cases, warpage could constitute a problem especially for GF containing formulas due to differential shrinkage along and across the oriented fiber direction. Impact strength is affected differently for GFs and talc. Depending on the level of impact strength of the matrix polymer used, GF can decrease, increase, or have a neutral effect. On the other hand, impact strength in talc containing PPs is most influenced by whether the talc used has high lamellarity or a coarse morphology/microcrystalline structure, as well as the level of dispersion.

HGMs, similar to reinforcing fillers, improve various properties including dimensional stability (sink, warpage, coefficient of thermal expansion (CTE)) but they typically reduce tensile, flex, and impact strength of the base polyolefin in their untreated form. With surface treatment of HGMs, tensile and flex strength can be improved in certain compound systems (e.g., aminosilane treatment in chemically coupled GF reinforced PP). The level of reduction is different depending on various factors. In the rest of this section, we will demonstrate how the properties of HGM containing TPO formulations can be recovered via careful selection of the amount of reinforcing fillers, impact modifiers, compatibilizers as well as the use of surface treated HGMs. We will look at each reinforcing filler system independently.

HGMs in GF Filled PP

GFs are widely used in TPO formulations and other polymers primarily to increase the strength and modulus of the polymer matrix phase. GFs also increase the impact strength of inherently low-impact strength polymers such as a homopolymer PP. The mechanism of impact strength increase with GFs is not related to impact absorption and cavitation, the accepted mechanistic theory for rubbers. Rather, GFs orient in the direction of injection molding and form a barrier to crack propagation as depicted in Figure 3.15. Figure 3.16 shows the effect of increasing GF loading from 0 wt% to 30 wt% on the impact strength of unfilled homopolymer PP. The impact strength almost triples from 26 J/m to 73 J/m when GF loading is increased from 0 wt% to 30 wt%. Note that the starting PP resin has low impact strength—26 J/m. We can also see in Figure 3.16 that the impact strength of 30 wt% GF containing PP reduces from 73 J/m to 34 J/m when we partially replace GFs with HGMs

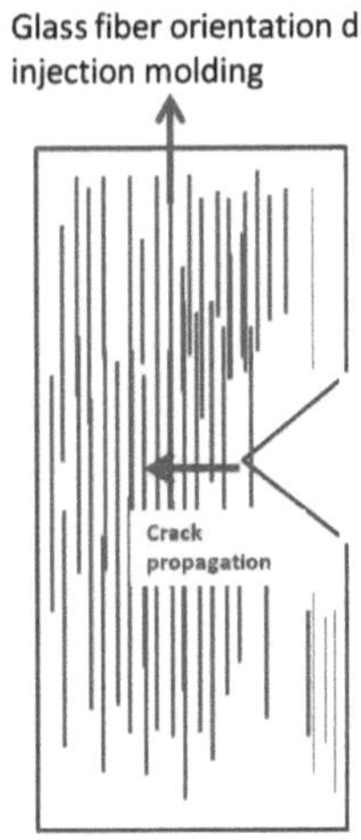

Figure 3.15 Schematics demonstrating oriented glass fibers forming a barrier to a crack propagating perpendicular to them.

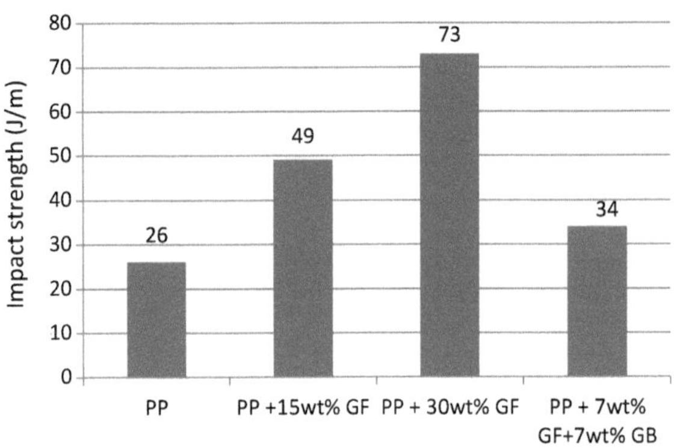

Figure 3.16 Impact increase in polypropylene (PP) with glass fibers and then decrease by partial hollow glass microsphere replacement.

(7 wt% GF and 7 wt% HGM instead of 30 wt% GF) such that the PP resin on a volume basis stays constant. This is due to the fact that GFs are in this case acting not only as a strength and modulus reinforcing filler but also as an impact strength increasing filler.

For reasons explained above, it is recommended that GF loading remains significantly unchanged when formulating with HGMs unless some mechanical strength decrease can be tolerated. Formulas 1 and 2 in Table 3.6 shows that density can be decreased significantly but comes at

Table 3.6 Physical Property Comparisons for Equal Vol% Loading of Hollow Glass Microsphere (HGM) versus GF in Polypropylene (PP)

	Formula 1		**Formula 2**	
	20 wt% GF		**8 wt% HGM (0.46, 16 KPSI)**	
Component	**Wt%**	**Vol%**	**Wt%**	**Vol%**
Homopolymer PP Albis	80	92.07	92	85.45
Glass fiber	20	7.93		
iM16K-HGM			8	14.55
Final	100	100	100	100
Density	1.035		0.853	
Tensile strength MPa (D-638)	76.9		29.8	
Tensile elongation % (D-638)	3.61		3.67	
Tensile modulus MPa (D-638)	3530		2397	
Flexural strength MPa (D-790)	98.6		56.5	
Flexural modulus MPa (D-790)	2730		1848	
Izod impact strength at room temperature J/m2 (D-256)	6380		2050	

the expense of significant loss in mechanical properties when GFs are completely removed and replaced with HGMs.

For proper formulation in GF filled polyolefin systems, it is recommended:

1. Not to change GF loading level significantly
2. to replace partially the resin with HGMs and compensate the drop in viscosity by using a higher flow resin partially
3. to use chemical coupling, that is, compatibilizers (e.g., maleated PP)
4. to use surface treated HGMs

Below we demonstrate a case study where weight reduction and mechanical properties have been optimized for a chemically coupled heat stabilized GF reinforced PP system by utilizing the above suggestions.

Case Study—Chemically Coupled GF Reinforced PP

For this study, HGM filled formulations were prepared by blending HGM masterbatch and GF filled polymer pellets at the injection molding machine. Table 3.7 shows the PP materials used in this study.

PP 30 wt% GF material was used as received in the molding process to create the standard control part. In order to prepare 31.5 wt% GF and 9 wt% HGM containing part, 70 parts by weight of P7-45FG-0790 BK711 was dry blended with 30 parts by weight of 3M iM16K HGM masterbatch and added to the hopper of the injection molding machine.

Table 3.7 Control Polypropylene (PP) 30 wt% GF Material Used in the Standard Part

P6-30FG-0600 BK711 (Asahi Kasei North America Plastics)	30 wt% glass filled, heat stabilized, chemically coupled injection moldable PP-H resin, 1.122 g/cc
P9900-H1165-B (Melt flow index 13) (Asahi Kasei North America plastics)	Unfilled, heat stabilized, injection moldable, PP-H resin with chemical coupling agent 0.900 g/cc
Hollow glass microspheres masterbatch in P9900-H1165-B	(both with silane treated and untreated) 30 wt% HGM (both silane treated and untreated) filled P9900-H1165-B 0.7258 g/cc
P7-45FG-0790 BK711 (Asahi Kasei North America plastics)	45% Glass filled, heat stabilized, chemically coupled injection moldable PP-H resin 1.29 g/cc

Molded Part Formulas—Mix Ratios of the HGM Masterbatch and as Received Materials

For this study, the molded part shape was ASTM test specimens. PP 30 wt% GF material was used as received in the molding process to create the standard control part. In order to prepare 31.5 wt% GF and 9 wt% HGM containing part, 70 parts by weight of P7-45FG-0790 BK711 was dry blended with 30 parts by weight of 3M iM16K HGM masterbatch and added to the hopper of the injection molding machine. Other ratios are shown in Table 3.8.

Mechanical Properties

Table 3.9 shows the mechanical properties and density reduction achieved with HGMs. A parenthesis next to a number indicates the value achieved when surface treated HGMs were used. If there is no parenthesis next to a number, it means that the difference in value was insignificant for untreated and surface treated HGMs and simply thc value achieved via untreated HGM is shown. Specific modulus values were determined by dividing the absolute modulus values by the density of the parts.

Table 3.8 Formulations (Mix Ratios) Used for Injection Molding of PP-GF and PP-GF-HGM Compounds

	Component	Density g/cc	Weight %	Volume %
30GF -Control	P6-30FG-0600 BK711	1.122	100.0	100.0
	Totals	**1.122**	**100.0**	**100.0**
31.5 GF 9.0 GB	P7-45FG-0790 BK711	1.2686	70.0	57.2
	P9900-H1165-B -GB30	0.7258	30.0	42.8
	Totals	**1.0361**	**100.000**	**100.000**
29.25 GF 10.5 GB	P7-45FG-0790 BK711	1.2686	65. 0	51.5
	P9900-H1165-B -GB30	0.7258	35.0	48.5
	Totals	**1.0054**	**100.000**	**100.000**
22.5 GF 15.0 GB	P7-45FG-0790 BK711	1.2686	50.0	36.4
	P9900-H1165-B -GB30	0.7258	50.0	63.6
	Totals	**1.0054**	**100.000**	**100.000**

Table 3.9 Hollow Glass Microspheres (HGM) Formulation in Chemically Coupled Heat Stabilized PP30F Formulation

Component		1		2		3		4		5	
		PP 15 GF		PP 30 GF		PP 31.5 GF 9 HGM		PP 29.5 GF 10.5 HGM		PP 22.5 GF 15 HGM	
		Wt%	Vol%	Wt%	Vol%	Wt%	Vol%	Wt%	Vol%	Wt%	Vol%
Polypropylene (PP)		85	94.1	70	86.8	59.5	69	60	67.7	62.5	64.1
Glass fiber		15	5.9	30	13.2	31.5	13	29.5	11.8	22.5	8.2
iM16K-HGM		-	-	-	-	9	18	10.5	20.5	15.0	27.7
Density		1.000		1.122		1.039		1.015		0.933	
% Reduction		−10.9		0.0		−7.4		−9.5		−16.8	
Tensile strength (MPa) D638		54.0		72.3		69 (78)		63.1 (74.6)		50 (61.0)	
Tensile modulus (MPa) D638	Specific modulus (MPa/g/cc)	2715	2715	4033	3594	5500	5294	5350	5271	4580	4909
Flexural modulus *(MPa) D790 (tangent 1%)	Specific modulus (MPa/g/cc)	3040	3040	4890	4360	5400	5200	5385	5300	4350	4660
Flexural strength (MPa) D790		80.0		107.9		106 (117)		112.5 (100)		81 (91)	
Room temperature Izod impact strength (J/m) D256		77		100		87		84		71	
Melt flow rate (230 °C, 2.16 kg)		4		4		3.8		3.9		3.8	

Mechanical properties of HGM containing parts in column 3, 4, and 5 are compared to column 2, the control part with 30 wt% GF. The properties for a 15 wt% GF formula are also shown for comparison in column 1 because lowering the amount of heavy filler is sometimes used as a means to decrease the density of compounds. One can see that lowering the amount of GF from 30 wt% to 15 wt% causes significant reduction in stiffness as measured by modulus and strength (compare column 1 and 2—tensile modulus (TM), tensile strength (TS), flexural modulus (FM), flexural strength (FS)) values. On the other hand, reduced density formulations containing HGMs in columns 3 and 4 display significant increase in modulus (both for treated HGMs and untreated), as well as retention of tensile and flexural strength for untreated HGM containing samples and increased tensile and flexural strength for samples containing surface treated HGMs. There is a slight decrease in the impact strength of the HGM containing samples but the impact level is still high for most materials used in this application. Small amounts of impact modifiers (less than 5 wt%) could help bring back the impact with little effect on the modulus.

It is also interesting to note that the viscosity, as measured by melt flow rate (MFR), is maintained although the resin content is significantly reduced (compare formula 2 at 86.8 vol% and formula 4 at 67.7 vol%). This is because a PP with a lower viscosity (higher MFR) was used to prepare the PP HGM masterbatch.

Similar results are obtained when the formulations shown in Table 3.9 are precompounded to the final composition rather than blended at the injection molding machine.

Similar results are obtained for long GF filled PPs as shown in Table 3.10.

HGMs in Talc Filled PP

Talc is a platy mineral which is commonly used in TPO formulations in order to increase the modulus, strength, and HDT of polyolefins. Although talc is not as reinforcing as GF, it enables good surface finish and dimensional stability to the parts, both of which could be an issue with GF filled TPO systems. Talc filled PPs are primarily used for interior automotive parts. Talc comes in many morphologies (microcrystalline, macrocrystalline), delamination levels (lamellar or coarse), and purities, which influence the final mechanical properties.

Coarse talc with low levels of lamellarity (reduced aspect ratio) reduces impact strength similar to HGMs in PP and therefore can be partially

Table 3.10 Hollow Glass Microsphere (HGM) Formulation in Chemically Coupled Heat Stabilized PP30LGF Formulation

Component		1		2		3		4	
		PP 30 GF Solid (Control)		PP 30 GF-7HGM-A Solid		PP 23 GF-5HGM-A Solid		PP 20 GF-8HGM-A Solid	
		Wt%	Vol%	Wt%	Vol%	Wt%	Vol%	Wt%	Vol%
Polypropylene (PP)		70.0	86.8	63	73.3	72	81.4	72	77.2
Long glass fiber		30.0	13.2	30	12.4	23	9.2	20	7.6
iM16K-HGM (surface treated)		-	-	7	14.4	5	9.4	8	15.1
Density		1.115		1.051		1.018		0.961	
% Density Reduction with respect to control 30 GF		-		5.8		8.6		13.8	
Tensile strength (MPa) D638		93.4		92.8		86.9		76.4	
Tensile elongation % D638		3.8		3.8		3.5		4.0	
Tensile modulus (MPa) D638	Specific modulus (MPa/g/cc)	5120	4592	5263	5007	5141	5050	4025	4188
Flexural modulus* (MPa) D790 (tangent 2%)	Specific modulus (MPa/g/cc)	3726	3341	4185	3982	3676	3609	2714	2926
Flexural strength (MPa) D790		131.7		132.9		123.8		110.0	
Room temperature Izod impact strength (J/m) D256		126		95		90		92	

replaced with HGMs. In contrast to GF formulations, there is no need to completely maintain the original level of talc reinforcement.

Table 3.11 shows that talc in a low-impact homopolymer PP (formula 1) can partially be replaced with small loadings of HGMs and the mechanical properties are significantly retained. When compatibilizers are used in such low impact formulas, properties such as tensile and impact strength improve further. It is important to note that impact strength increases with maleated PP compatibilizers can only be achieved in low-impact homopolymer PPs. When higher impact copolymer grades are used, the use of compatibilizer alone does not recover the impact strength. For those higher impact copolymers, it is recommended that impact modifiers along with compatibilizers are used. Tables 3.12 and 3.13 are examples of higher impact grades of PP copolymers utilizing lower loadings of talc. In these systems, it becomes more difficult to replace the already low levels of talc and still achieve 10% density reduction along with well-maintained mechanical properties. In such polymer systems, the simultaneous use of impact modifiers and compatibilizers are inevitable if mechanical property retention is required to high extent. We will elaborate on the simultaneous use impact modifiers and compatibilizers in the upcoming section entitled HGMs in unfilled polyolefins.

HGMs in Unfilled Polyolefins

In this section, we will look at the effect of impact modifiers and compatibilizers on the properties of polyolefins, primarily for PP. Table 3.14 shows the impact modifier and compatibilizer as well as the HGM grades used in these studies.

When using HGMs in unfilled PP resins with medium to high impact strength such as in copolymers or rubber filled grades, it is imperative that compatibilizers and impact modifiers are used simultaneously. One can see in Formula 1 of Table 3.15, the impact strength of PP copolymer reduces significantly from about 205 J/m to 47 J/m with 14 wt% HGMs. The reduction is also seen in tensile and flex strength while the modulus increases. When an impact modifier (polyolefin elastomer) is added at 17 wt%, the impact strength more than doubles to 120 J/m but does not recover the original impact strength level of the PP. In addition, the tensile strength further reduces due to the soft elastomeric nature of the impact modifier. When a small amount of compatibilizer is added in addition to the impact modifier, the impact strength surprisingly increases to 273 J/m

Table 3.11 Mechanical Properties of Talc Reinforced Polypropylene (PP) Homopolymer in the Presence of Hollow Glass Microspheres (HGMs)

	Formula 1		Formula 2		Formula 3	
	PP-T20		PP-T10 HGM 4		PP-T10 HGM 4 -MAPP	
Component	**Wt%**	**Vol%**	**Wt%**	**Vol%**	**Wt%**	**Vol%**
PP homopolymer	80	92.5	86	88.6	83	85.5
HGM (0.46 g/cc)			4	8	4	8
Talc	20	7.5	10	3.4	10	3.4
MAPP compatibilizer					3	3.1
Final	100	100	100	100	100	100
Density	1.046		0.942		0.943	
Tensile strength @ room temperature (RT) (MPa)	31.7		27.0		32.7	
Tensile strength @ 90 °C (MPa)	12.5		11.4		13.5	
Tensile elongation (%)	10		40		12	
Tensile modulus @ RT (MPa)	2110		1900		1835	
Tensile modulus @ 90 °C (MPa)	270		265		250	
Flexural strength (MPa)	49		45		50	
Flexural modulus @1% secant (MPa)	1650		1620		1620	
Izod impact strength at RT (J/m)	32		28		39	

MAPP = maleic anhydride modified polypropylene.

Table 3.12 Mechanical Properties of Talc Reinforced Polypropylene (PP) Copolymer in the Presence of Hollow Glass Microspheres (HGMs)

	Formula 1		Formula 4		Formula 5	
	PP-Talc		PP–Talc- HGM EN8407-PB3200		PP–Talc- HGM EN8407-PB3200	
Component	**Wt%**	**Vol%**	**Wt%**	**Vol%**	**Wt%**	**Vol%**
PP-BASE PP	95	98.3	70.3	64.6	76.95	74.6
HGM (0.46 g/cc)	-	-	14	22.6	7	11.9
Talc	5	1.7	3.7	1.1	4.05	1.3
PB3200 maleated compatibilizer	-	-	4	3.7	4	3.8
Engage 8407 impact modifier	-	-	8	7.7	8	8.1
Density	0.931		0.834		0.876	
% Reduction in density	-		10.4		5.9	
Tensile strength @ room temperature (RT) (MPa)	21.1		17.3		18.0	
Tensile modulus @RT (MPa)	1370		1323		1221	
Tensile elongation @ RT (MPa)	46.4		11.0		22.0	
Flexural strength (MPa)	32.3		29.6		28.7	
Flexural modulus @1% secant (MPa)	1285		1241		1134	
impact strength at RT (J/m) notched D256	112.1		110		120	
Melt flow index (230 °C 2.16 kg)	25.5		10.0		15.8	

Table 3.13 Mechanical Properties of Talc Reinforced Polypropylene (PP) Copolymer in the Presence of Hollow Glass Microspheres (HGMs)

	Formula 1		Formula 6	
	14 Talc		10 Talc-7HGM 12.5 Impact MOD	
Component	**Wt%**	**Vol%**	**Wt%**	**Vol%**
Hostacom base PP	86	95	65.8	64.7
HGM-iM16K	-	-	7	13.2
PB3200	-	-	3.9	4.2
Engage 8407	-	-	12.6	14.2
Talc	14	5	10.7	3.7
Density	1.000		0.914	
% Reduction in density	-		8.5	
Tensile strength @ room temperature (RT) (MPa)	20.4		16.6	
Tensile modulus @ RT (MPa)	1554		1042	
Tensile elongation @ RT (MPa)	44.0		38.0	
Flexural modulus @ 1% secant (MPa)	1327		1010	
impact strength at RT (J/m) notched D256	268		241	
Impact strength at RT (J/m) unnotched D256	1080		1165	
Melt flow index (230 °C 2.16 kg)	36.8		12.6	

Table 3.14 Formula Ingredients Used in This Section

Main Resin	Impact Polypropylene (PP) Copolymer
Compatibilizer	Maleic anhydride modified PP-H under the trade name POLYBOND® 3200 available from Addivant. Melt flow rate (MFR) (190C/2.16 kg) 115 g/10 min. 0.8–1.2 % maleic anhydride content
Impact modifier	Polyolefin elastomer Engage® 8137 with a MFR (190C/2.16 kg) 13 g/10 min from Dow Chemical Company
Impact modifier	Polyolefin elastomer Engage® 8407 with a MFR (190C/2.16 kg) 30 g/10 min from Dow Chemical Company
Hollow glass microsphere	3M™ iM16K with 16,000 psi crush strength, 20 micron average diameter, and 0.46 g/cc true density

exceeding that of the unfilled PP. In addition, the tensile strength also increases compared to the impact modifier only containing formula 3. In the formulations where the impact modifier and compatibilizers are used, it is also shown that using an impact modifier with a low MFI increases the impact strength (Formula 6) more efficiently compared to those with high MFI (Formula 4). One can see that the impact strength at the same impact modifier loading is twice as that of the unfilled PP control (Formula 1).

Same behavior is observed in a PP copolymer with a high MFI of 50 g/10 min (@230 °C 2.16 kg) in Formula 1 of Table 3.16. Formula 2 corresponds to the 14 wt% (22 vol%) HGM loading in this impact copolymer PP. One can notice that the density of the unfilled PP reduces from 0.900 g/cc to 0.817 g/cc. Modulus, on the other hand, increases 36%. However, reduction is observed for the tensile, flexural, and impact strength of the compound with the impact strength exhibiting the most considerable drop. In order to compensate for the reduction in impact strength, Formula 3 utilizes an impact modifier, that is, Engage 8407 at 17 wt%. Although the impact strength recovers considerably, the flexural and tensile strength is further decreased due to the soft nature (low modulus and strength) of the impact modifier. In Formula 4, we add 4 wt% compatibilizer on top of the 17 wt% impact modifier. One can see that the

Table 3.15 Mechanical Properties of Hollow Glass Microsphere (HGM) Filled Polypropylene (PP) Copolymer in the presence of Impact Modifier and Compatibilizer

	Formula 1		Formula 2		Formula 3		Formula 4		Formula 5		Formula 6	
	PP Control		PP-HGM14		PP-HGM14 EN 8407		PP-HGM14-PB3200 EN 8407		PP-HGM14-PB3200 EN 8137		PP-HGM14-PB3200 EN 8100	
Component	**Wt%**	**Vol%**	**Wt%**	**Vol%**	**Wt%**	**Vol%**	**Wt%**	**Vol%**	**Wt%**	**Vol%**	**Wt%**	**Vol%**
PP 4208 copolymer	100	100	90	78	69	62	65	58.3	65	58.3	65	58.3
HGM-iM16K (0.46 g/cc)	-	-	14	22	14	22	14	22.0	14	22.0	14	22.0
PB3200	-	-	-	-	-	-	4	3.6	4	3.6	4	3.6
Engage 8100 (Melt flow index (MFI):1)	-	-	-	-	-	-	-	-	-	-	17	16.1
Engage 8137(MFI:15)	-	-	-	-	-	-	-	-	17	16.1	-	-
Engage 8407(MFI:30)	-	-	-	-	17	16	17	16.1	-	-	-	-
Density (g/cc)	0.90		0.814		0.813		0.815		0.815		0.815	
Tensile strength @ room temperature (RT) (MPa)	21.6		13.0		10.6		15.2		15.0		15.6	
Tensile modulus @ RT (MPa)	906		1186		778		866		810		810	
Tensile elongation @ RT %	Limit		134		Limit		60		80		80	
Flexural strength (MPa)	27.9		22.7		15.7		21.7		21.1		21.6	
Flexural modulus @ 1% secant (MPa)	915		1209		812		883		848		853	
Impact strength at RT (J/m) notched D256	205		47		120		273		311		397	
Impact strength at RT (J/m) unnotched D256	1235		755		787		1030		1017		925	
MFI (230 °C 2.16 kg)	8.7		5.3		4.6		5.3		5.0		4.1	

Table 3.16 Impact Copolymer Polypropylene (PP) Compounds with Hollow Glass Microspheres (HGMs), Impact Modifiers, and Compatibilizers

	Formula 1		Formula 2		Formula 3		Formula 4	
	Wt%	Vol%	Wt%	Vol%	Wt%	Vol%	Wt%	Vol%
PP 4150H copolymer	100	100	90	78	69	62	65	58.3
HGM-iM16K (0.46 g/cc)	-	-	14	22	14	22	14	22.0
PB3200	-	-	-	-	-	-	4	3.6
Engage 8407 (MFI:30)	-	-	-	-	17	16	17	16.1
Density (g/cc)	0.90		0.817		0.818		0.812	
Tensile strength @ room temperature (RT) (MPa)	18.2		11.1		8.9		12.6	
Tensile modulus @ RT (MPa)	917		1245		802		782	
Tensile elongation @RT %	30		13		52		23	
Flexural strength (MPa)	26.0		20.7		15.0		19.2	
Flexural modulus @ 1% secant (MPa)	866		1119		761		788	
impact strength at RT (J/m) notched D256	128		41		125		260	
Melt flow index (230 °C 2.16 kg)	55		27		22		17.6	

impact strength improves significantly upon the addition of compatibilizer from 125 J/m (Formula 3) to 260 J/m (Formula 4). More interesting is the simultaneous increase in flexural and tensile strength (compare Formula 3 and 4) which typically reduces as impact increases.

Using these two examples in Tables 3.15 and 3.16, we can conclude that the maleic anhydride grafted PP compatibilizer improves the efficiency of the polyolefin elastomer impact modifier in HGM filled PP compounds. This behavior is summarized in Figure 3.17. It is also interesting to note that the simultaneous use of impact modifier and compatibilizer increases the impact strength of the HGM only compound (Formula 2) by 530% from 41 J/m to 260 J/m while maintaining tensile and flexural strength.

Although not shown here, it is important to mention that this behavior, that is, considerable increase in impact strength via the combination of compatibilizer and impact modifier, is not observed when there is no HGM in the formula. In other words, in impact copolymer polypropylene, the impact strength can be increased via the use of impact modifiers but it does not increase any further with the addition of compatibilizers when there is no HGM in the formula of the compound.

Since the combination of impact modifier at 17 wt% with 4 wt% compatibilizer improves impact strength considerably to more than double the amount of the unfilled control PP, next we determine at what loading content the impact modifier amount is enough to match the room temperature impact strength of the unfilled impact copolymer polypropylene. We can see in Table 3.17 and Figure 3.18 that about 7 wt% impact modifier with 4 wt% compatibilizer can recover unfilled PP

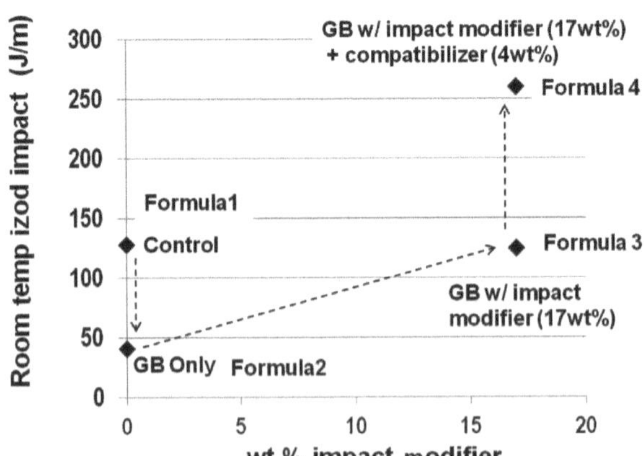

Figure 3.17 Behavior of impact strength with hollow glass microspheres (HGMs), HGM/impact modifier, and HGM/impact modifier/compatibilizer.

Table 3.17 Effect of Impact Modifier Amount (@ 4% Compatibilizer)

	Formula 1		Formula 2		Formula 3		Formula 4	
	Wt%	**Vol%**	**Wt%**	**Vol%**	**Wt%**	**Vol%**	**Wt%**	**Vol%**
Polypropylene (PP) 4150H copolymer	100	100	90	78	77.7	70.6	73.8	66.9
Hollow glass microsphere-iM16K (0.46 g/cc)	-	-	14	22	14.0	21.7	14.0	21.7
PB3200	-	-	-	-	4.0	3.6	4.0	3.6
Engage 8407 (melt flow index (MFI):30)	-	-	-	-	4.3	4.1	8.2	7.7
Density (g/cc)	0.90		0.817		0.812		0.811	
Tensile strength @ room temperature (RT) (MPa)	18.2		11.1		16.6		14.9	
Tensile modulus (MPa)	917		1245		1086		1040	
Flexural strength (MPa)	26.0		20.7		25.5		23.6	
Flexural modulus @1% secant (MPa)	866		1119		990		937	
impact strength at RT (J/m) notched D256	128		41		83		142	
MFI (230 °C 2.16 kg)	55		27		25.8		21.8	

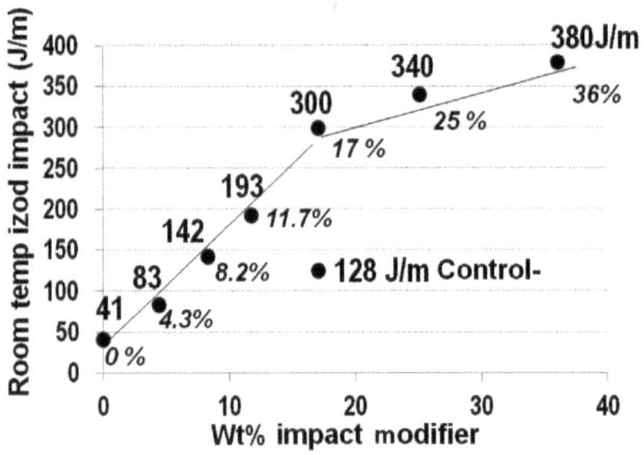

Figure 3.18 Effect of Impact modifier amount on impact strength (all formulas contain 4 wt% compatibilizer and 14 wt% hollow glass microspheres (HGMs) except the first data point which contains 14 wt% HGM only and no compatibilizer).

impact strength while maintaining its flexural strength and modulus levels. Figure 3.18 also shows that the increase in impact strength is pretty linear until 17% after which the rate of impact strength increase decreases.

Figure 3.19 shows the effect of compatibilizer loading on the impact strength at 17 wt% impact modifier loading. One can see that the impact

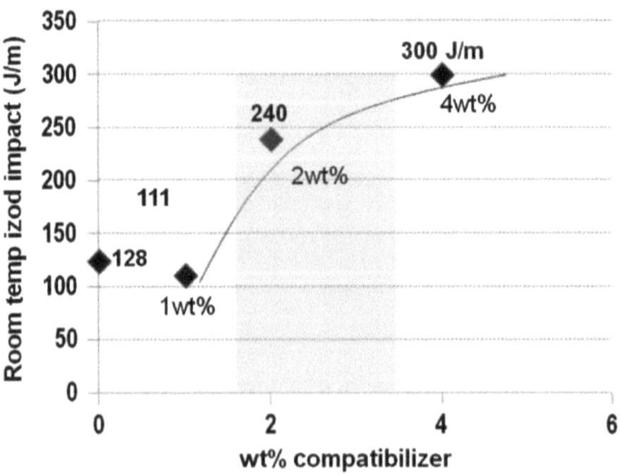

Figure 3.19 Effect of compatibilizer amount on impact strength (all formulas contain 17 wt% impact modifier except the control unfilled polypropylene at 128 J/m impact strength).

strength does not increase below and at 1 wt% compatibilizer. In fact, there is a small drop in impact strength at this concentration range. Above this concentration, impact strength starts to increase substantially and levels off above 4 wt%.

Table 3.18 shows a high impact copolymer polypropylene manufactured using the LyondellBasell's proprietary Catalloy process technology. Similar to that shown in Table 3.6, very high impact strength levels can be achieved via the combined use of impact modifiers and compatibilizer.

Similar mechanical property changes are observed for PE with HGMs and all these properties can be improved with the use of impact modifiers and compatibilizers. It is important, however, that maleated PE compatibilizers be used for PE systems. In the selection of impact modifiers, the use of higher melt flow impact modifiers prevent reduction in the viscosity of the final composite systems but could be less efficient in improving impact. Table 3.19 shows that the use of impact modifiers (Engage 8137 and Engage 8100) alone does little in increasing impact and reduces tensile strength further. However, when a small amount of compatibilizer is added, the impact strength increases almost 400% while increasing tensile strength simultaneously. Higher viscosity polyolefin elastomer (Engage 8100) increases impact strength more efficiently but reduces the MFI of the entire system which may not be desirable for injection molding process.

Table 3.20 shows that the choice of compatibilizer chemistry is also important. Choosing a maleated polymer with a backbone that is not compatible (maleated PP) with the main matrix (PE) system does not bring about the same effect of increasing the efficacy of the impact modifier and increasing tensile strength simultaneously (compare Formula 2 and 3). Similarly Table 3.21 shows that maleated PE does not bring about the benefits of increasing the efficiency of the impact modifier in a PP/HGM composite.

One polyolefin system that does not follow the general rules of mechanical property changes with HGMs is linear low density polyethylene (LLDPE). Table 3.22 shows that the impact strength, modulus, and flexural strength all increase with increased loadings of HGMs without resorting to impact modifiers and compatibilizers.

HGMs in PA

PA, also commonly called "Nylon," are used in significant quantities in automotive applications due to their high toughness, dimensional and thermal stability, high continuous use temperatures, and reasonable cost.

Table 3.18 Impact Copolymer Polypropylene (PP) Compounds with Hollow Glass Microspheres (HGMs), Impact Modifiers, and Compatibilizers

	Formula 1		Formula 2		Formula 3	
	Wt%	Vol%	Wt%	Vol%	Wt%	Vol%
Hifax CA387A	100	100	91	79.5	66	59.4
HGM-iM16K (0.46 g/cc)	-	-	13	20.5	13	21.1
PB3200	-	-	-	-	4	3.6
Hifax CA138A	-	-	-	-	17	15.9
Density (g/cc)	0.90		0.824		0.824	
Tensile strength @ room temperature (RT) (MPa)	15.9		10.0		13.6	
Tensile modulus (MPa)	846		993		865	
Tensile elongation %	33		34		30	
Flexural strength (MPa)	19.9		14.9		19.4	
Flexural modulus @ 1% secant (MPa)	747		780		802	
impact strength at RT (J/m) notched D256	660		153		316	
Melt flow index (230 °C 2.16 kg)	18.9		8.6		5.8	

Table 3.19 Effect of Hollow Glass Microspheres (HGMs) on Mechanical Properties of High Density Polyethylene (HDPE) in the Presence of Impact Modifiers and Compatibilizer

	HDPE Contro		HDPE-HGM12 EN8100		HDPE-HGM12 PB-EN8100		HDPE-HGM12 EN8137		HDPE-HGM12 PB-EN8137	
	Wt%	Vol%	Wt%	Vol%	Wt%	Vol%	Wt%	Vol%	Wt%	Vol%
HDPE	100	100	70.5	62.5	66.2	58.7	70.5	62.5	66.2	58.7
HGM-iM16K	-	-	12.0	20.4	12.0	20.4	12.0	20.4	12.0	20.4
Engage 8137 (melt flow index (MFI): 15)	-	-	-	-	-	-	17.5	17.1	17.7	17.2
Engage 8100 (MFI: 1)	-	-	17.5	17.1	17.7	17.2	-	-	-	-
Polybond 3009 (maleated PE)	-	-	-	-	4.1	3.7	-	-	4.1	3.7
Density (g/cc)	0.959		0.854		0.857		0.851		0.852	
% Reduction in density	-		11.0		10.6		11.2		11.2	
Tensile strength @ room temperature (RT) (MPa)	23.2		13.6		18.0		13.1		18.0	
Tensile elongation (%)	Limit		Limit		36		Limit		40	
Tensile modulus@ RT (MPa)	870		840		800		765		858	
Flexural strength (MPa)	24.9		18.2		20.0		18.0		19.7	
Flexural modulus @ 2% secant (MPa)	694		570		560		570		562	
Flexural modulus @ 1% secant (MPa)	810		700		680		686		667	
Izod impact strength at RT (J/m)-notched D256	91		63.5		362		65.0		275	
Izod impact strength at RT (J/m)-unnotched	1200		882		964		888		900	
MFI (190 °C 2.16 kg)	5		2.6		2.65		4.5		4.3	

Table 3.20 Effect of Compatibilizer Backbone Chemistry on Mechanical Properties of High Density Polyethylene (HDPE)/ Hollow Glass Microsphere (HGM) Composite

	Formula 1		Formula 2		Formula 3	
	HDPE Control		HDPE-HGM12 PB3009-EN8137		HDPE-HGM12 PB3200-EN8407	
	Wt%	Vol%	Wt%	Vol%	Wt%	Vol%
HDPE	100	100	66.2	58.7	66.2	58.7
HGM-iM16K (0.46 g/cc)	-	-	12.0	20.4	12.0	20.4
Engage 8137	-	-	17.7	17.2	-	-
Polybond 3200 (maleated polypropylene)	-	-	-	-	17.7	17.2
Polybond 3009 (maleated polyethylene)	-	-	4.1	3.7	4.1	3.7
Density (g/cc)	0.959		0.852		0.834	
Tensile strength @ room temperature (RT) (MPa)	23.2		18.0		13.0	
Tensile elongation (%)	Limit		40		Limit	
Tensile modulus @ RT (MPa)	870		858		700	
Flexural strength (MPa)	24.9		19.7		16.8	
Flexural modulus @ 2% secant (MPa)	694		562		487	
Flexural modulus @1% secant (MPa)	810		667		596	
Izod impact strength at RT (J/m)-notched D256	91		275		43	
Izod impact strength at RT (J/m)-unnotched	1200		900		610	
Melt flow index (190 °C 2.16 kg)	5 (7.5)		4.3		3.5	

Table 3.21 Effect of Compatibilizer Backbone Chemistry on Mechanical Properties of Polypropylene (PP)/Hollow Glass Microsphere (HGM) Composite

	Formula 1		Formula 2		Formula 3		Formula 4	
	PP		PP-HGM14		PP-HGM14-PB3200 EN 8137-17 wt% rubber		PP-HGM14-PB3009 EN 8137-17 wt% rubber	
Component	Wt%	Vol%	Wt%	Vol%	Wt%	Vol%	Wt%	Vol%
PP 4150H	100	100	90	78	65	58.3	64.7	58.2
HGM-iM16K (0.46 g/cc)	-	-	14	22	14.0	22.0	14.0	22.2
Polybond 3200 (maleated PP)	-	-	-	-	4.0	3.6	-	-
Polybond 3009 (maleated polyethylene)	-	-	-	-	-	-	4.0	3.4
Engage 8137	-	-	-	-	17	16.1	17.3	16.2
Density (g/cc)	0.90		0.817		0.823		0.813	
Tensile strength @ room temperature (RT) (MPa)	18.2		11.1		12.8		9.6	
Tensile modulus @ RT (MPa)	917		1245		780		590	
Tensile elongation @ RT %	30		13		22		43	
Flexural strength (MPa)	26.0		20.7		19.0		14.4	
Flexural modulus @1% secant (MPa)	866		1119		767		543	
impact strength at RT (J/m) notched D256	128		41		300		89	
Melt flow index (230 °C 2.16 kg)	55		27		18.5		18.4	

Table 3.22 Mechanical Property Changes in Linear Low Density Polyethylene (LLDPE) with Hollow Glass Microspheres (HGMs)

	Formula 1		Formula 2		Formula 3	
	LLDPE		LLDPE -15HGM		LLDPE -25HGM	
Component	**Wt%**	**Vol%**	**Wt%**	**Vol%**	**Wt%**	**Vol%**
LLDPE (exceed 3518)	100	100	82.5	72	77	65
HGM-iM16K (0.46 g/cc)	-	-	17.5	28	23	35
Final	100	100	100	100	100	100
Density (g/cc)	0.918		0.802		0.775	
Tensile strength at yield @ room temperature (RT) (MPa) D638	10		10		9.0	
Tensile elongation @ RT (%) D415	650		510		400	
Tensile modulus @ RT (MPa) D638	145		265		300	
Flexural strength (MPa) D790	9.6		11.7		13.1	
Flexural modulus (MPa) D790	185		314		366	
Notched Izod impact strength at RT (J/m) D256	370		395		445	
Melt flow rate 190 °C, 2.16 g/10 min	3.8		1.3		0.85	
Durometer shore A	92.9		95		95.6	

Common applications include engine covers, battery trays, grill opening reinforcements, and front end modules. Typically automotive applications employ PA6 or PA66 base resin that are reinforced with GF and/or mineral. The GF is generally coated with a polymer-specific sizing agent to increase the bond to the base resin. Similar to the previous sections, direct comparisons of physical properties are made of control systems—PA with typical GF loading—compared to HGM containing systems. All physical properties were measured on dry as molded samples that were kept in a low relative humidity chamber for at least 48 h for cooling and conditioning prior to testing.

Commercially available, high strength (16KPSI, 18KPSI, and 27KPSI isostatic crush strength), low density (0.46 and 0.60 g/cc) HGMs (3M™ HGMs iM16K, S60HS, and iM30K) were selected for comparative formulations. A commercially available, injection molding grade of PA66 (also referred to as Nylon 66) was obtained from E.I. DuPont de Nemours Company under the trade name Zytel® 101LNC010. In cases where GFs were not already present in the base resin, PPG™ Chopvantage 3540 for PAs was used.

Flexural strength and flexural moduli were determined according to the ASTM D790. Notched Izod impact properties were determined according to ASTM D252. Tensile mechanical properties were determined according to the ASTM D638. The density of the injection molded parts was determined using a helium gas pycnometer.

There are two main directions that can be taken in compounding commercially available resin systems with HGMs. One tactic is to use a material that already contains other reinforcing fillers such as GF and then add HGMs "on top" of the existing formula through an additional compounding step or by adding the HGMs via a masterbatch during the injection molding process. This will, in effect, dilute the resin and GF content as the HGM content (and masterbatch carrier resin) increases. This is an important formulation parameter to consider. Often the dilution of the fiber content will reduce some of the reinforcing physical properties beyond what may be acceptable for the application. Sometimes it may be appropriate to add back some GF content to retain these properties but also add higher levels of HGMs to reduce the overall density of the composite. The other tactic is to make the optimized resin formulation from scratch in a single compounding process by starting with an unfilled resin and adding the discrete amounts of fiber and HGM in sequence, as described above. The results in Tables 3.23, 3.24, and 3.25 show property changes for the "on top" method. At the top of each table the PA, GF, and HGM contents are identified in wt% and vol%. It is evident from the "Density" data how the

Table 3.23 Comparison of Polyamide 66 (PA66) + 10 wt% Glass Fiber (GF) to Hollow Glass Microsphere (HGM) Containing Formulations: HGM Added "on Top"

	1		2		3		4		5	
	PA66 10 GF		PA66 10 GF 5HGM		PA66 9 GF 10HGM		PA66 8 GF 15HGM		PA66 8F 20HGM	
Component	**Wt%**	**Vol%**	**Wt%**	**Vol%**	**Wt%**	**Vol%**	**Wt%**	**Vol%**	**Wt%**	**Vol%**
PA66	90	95.3	85	85.9	81	77.9	77	70.6	72	63.4
GF	10	4.7	10	4.6	9	3.8	8	3.2	8	3.1
HGM	-		5	9.5	10	18.3	15	26.2	20	33.5
Density (g/cc)	1.21		1.15		1.10		1.05		1.0	
Tensile strength at yield (MPa) D638	73		82		80		77		70	
Tensile elongation at room temperature (RT) (%) D638	1.6		1.9		2		2		1.9	
Tensile Modulus (GPa) D638	5.5		6.3		5.4		6.2		6.1	
Flexural strength (MPa) D790	147		140		136		128		116	
Flexural modulus (MPa) D790	4495		4428		4706		4800		4997	
RT Izod impact strength (J/cm) D256	0.6		0.7		0.6		0.6		0.5	
Unnotched Izod impact strength (J/cm) D256	2.8		2.5		2.5		2.3		2.4	

Table 3.24 Comparison of Polyamide 66 (PA66) + 20 wt% Glass Fiber (GF) to Hollow Glass Microsphere (HGM) Containing Formulations; HGM Added "on Top"

	1		2		3		4		5	
	PA66 20 GF		**PA66 19 GF 5HGM**		**PA66 18 GF 10HGM**		**PA66 17 GF 15HGM**		**PA66 16 GF 20HGM**	
Component	**Wt%**	**Vol%**	**Wt%**	**Vol%**	**Wt%**	**Vol%**	**Wt%**	**Vol%**	**Wt%**	**Vol%**
PA66	80	90	76	80.8	72	72.8	68	65.4	64	56.7
GF	20	10	19	9	18	8	17	7.2	16	6.5
HGM	-	-	5	10.2	10	19.2	15	27.4	20	34.8
Density (g/cc)	1.27		1.22		1.18		1.4		1.1	
Tensile strength at yield (MPa) D638	108		103		118		113		109	
Tensile elongation at room temperature (RT) (%) D638	1.6		1.9		2		2		1.9	
Tensile modulus (GPa) D638	7.8		8.4		8.4		7.8		8.7	
Flexural strength (MPa) D790	196		189		184		174		160	
Flexural modulus (MPa) D790	6060		6339		6244		6868		6965	
RT Izod impact strength (J/cm) D256	0.8		0.5		0.5		0.8		0.9	
Unnotched Izod impact strength (J/cm) D256	3.9		4.3		4		4.2		4.5	

Table 3.25 Comparison of Polyamide 66 (PA66) + 33 wt% Glass Fiber (GF) to Hollow Glass Microsphere (HGM) Containing Formulations; HGM Added "on Top"

	1		2		3		4		5	
	PA66 33 GF		**PA66 31 GF 5HGM**		**PA66 30 GF 10HGM**		**PA66 28 GF 15HGM**		**PA66 27 GF 20HGM**	
Component	**Wt%**	**Vol%**	**Wt%**	**Vol%**	**Wt%**	**Vol%**	**Wt%**	**Vol%**	**Wt%**	**Vol%**
PA66	67	82	64	73.3	60	65	57	58.2	53	52.2
GF	33	18	31	10.9	30	14.4	28	12.6	27	11.1
HGM	-	-	5	15.7	10	20.6	15	29.2	20	36.7
Density (g/cc)	1.39		1.33		1.28		1.23		1.19	
Tensile strength at yield (MPa) D638	177		170		165		142		132	
Tensile elongation at room temperature (RT) (%) D638	2.2		2.3		2		2.1		2.1	
Tensile modulus (GPa) D638	12		11.5		11		10.7		9	
Flexural strength (MPa) D790	196		189		184		174		160	
Flexural modulus (MPa) D790	9266		9602		9000		9708		9545	
RT Izod impact strength (J/cm) D256	1.1		1.4		0.9		1.1		1.1	
Unnotched Izod impact strength (J/cm) D256	6.3		7		7.4		6.8		5.8	

increasing addition of the 0.6 g/cc, 18KPSI HGM (3M HGM S60HS) significantly reduces the density of the injection molded parts. Formulations sets such as these are useful in determining an optimized formulation. If the current resin is a 10, 20, or 33 wt% GF filled system, subsequent additions of HGMs can show the point at which the strength to density is optimized for a specific application. In these formulations, there tends to be an optimal point, where density reduction is high and physical properties are more or less maintained. As the concentration increases beyond this level the system behaves more glass-like, becoming stiffer and losing some tensile strength and impact. For instance, the formulation with the 10 wt% GF and 90 wt% Nylon 66 has a final part density of 1.21 g/cc with a tensile strength of 73 MPa, tensile modulus of 5.47 GPa, flexural strength of 147 MPa, flexural modulus of 4495, unnotched Impact of 2.8 J/cm, and notched Izod impact of 0.6 J/cm. Yet Formulas 2 and 3 containing 5 wt% and 10 wt% of the HGM, with 5% and 8.3% reduction in weight are almost identical with respect to physical properties.

Other interesting comparisons can be made. For example, to double all of the mechanical properties of the 10% GF base formulation, the fiber content has to be increased from 10 wt% to 33 wt% as shown by comparing Tables 3.23–3.25. Formulas 4 and 5 in Table 3.25 have essentially the same density as Formula 1 in Table 3.23 but roughly double the tensile strength, modulus, and impact strength. When such a high amount of fiber is present many of the properties can be retained even with a fairly high loading of HGM.

In the next series of data tables, the resin systems are formulated from scratch and the GF level is maintained as the resin is depleted to make room for the HGMs. In some of the tables, there are comparisons of a 0.6 g/cc, 27KPSI HGM to a 0.46 g/cc, 16KPSI HGM. These comparisons are formulated to achieve a similar final density for the formulation. The difference in HGM starting density and the resulting volume loss due to breakage of the different strength materials demonstrates one of the formulation keys stated above—find the lowest density HGM that survives the process. This yields the lowest final part density, the best set of physical properties and generally the lowest system cost since HGMs tend to be sold by weight, not volume.

A 10 wt% GF filled formulation with 4.7 wt% of the 0.46 g/cc HGM (3M™ HGM iM16K) provides similar tensile strength, elongation, tensile and flexural modulus to 10 wt% GF with 6 wt% of the 0.6 g/cc HGM (3M™ HGM iM30K). Also the physical properties are similar to the system without HGMs but the density is reduced 7.5%. Doubling the 0.46 g/cc HGM wt% has minimal further reduction in physical properties and 14% density reduction (Table 3.26).

Table 3.26 Comparison of Polyamide 66 (PA66) + 10 wt% Glass Fiber (GF) to Hollow Glass Microsphere Containing Formulations; Glass Fiber Level Maintained

	Formula 1		**Formula 2**		**Formula 3**		**Formula 4**	
	PA66 Base Resin + 10 wt% Glass Fiber		**PA66 Base Resin + 10wt% Glass Fiber + 6 wt% iM30K**		**PA66 Base Resin + 10 wt% Glass Fiber + 4.7 wt% iM16K**		**PA66 Base Resin + 10 wt% Glass Fiber + 10 wt% iM16K**	
Component	**Wt%**	**vol%**	**Wt%**	**vol%**	**Wt%**	**vol%**	**Wt%**	**vol%**
PA66	90	95.3	84	83.9	85.3	83.9	80	72.9
GF	10	4.7	10	4.5	10	4.4	10	4.1
HGM-iM30K			6.0	11.6				
HGM-iM16K					4.7	11.7	10	23
GB/GF vol ratio	-		2.6		2.7		5.6	
Final %	100	100	100	100	100	100	100	100
Vol% glass	4.7		16.1		16.1		27.1	
Density (g/cc)	1.21		1.14		1.12		1.04	
Tensile strength (MPa)	102		106		102		94	
Tensile elongation (%)	5		5.7		5.2		4.7	
Tensile modulus (MPa)	2565		2661		2524		2643	
Flexural modulus (MPa)	3924		4276		4037		4204	
Room temperature Izod impact strength (kJ/m^2)	5.1		3.2		3.3		2.9	

A 20 wt% GF filled system with 3% 0.46 g/cc HGM has equivalent physical properties to 20 wt% GF filled Nylon 66% and 4% density reduction. Again the difference in wt% needed to achieve a 1.23 g/cc compound density is evident comparing Formulas 2 and 3 in Table 3.27.

Table 3.28 shows a 30 wt% GF filled system with 5% 0.46 g/cc HGM has equivalent physical properties (except for a slight—7.5%—drop in tensile strength) to 30 wt% GF filled PA66 and 9% density reduction. Also, comparing Formula 3 in Table 3.2 to Formula 1 in Table 3.3, at the same volume % of glass and 10% density reduction, the HGM containing formulation provides 70% of impact strength, 80% of tensile strength and tensile modulus, and 90% of flexural modulus.

Figure 3.20 shows the effect of coating the HGMs with a coupling agent. The physical properties are normalized to 100%. There are two 0.6 g/cc HGMs having different strengths and slightly different particle size distributions. Comparing the uncoated (solid bar) versus the coated (textured bar) HGM for a specific grade, to the base resin property, clearly shows the benefits of coupling agent for improving tensile and flexural strength as well as elongation.

There are several other benefits of adding HGMs into thermoplastics compounding formulations. These can be outlined as reduced cycle time in thick molded parts, decreased mold shrinkage and part warpage, and reduced CTE as discovered during numerous laboratory and plant trials. Figure 3.21 shows the effect of HGM loading on the coefficient of linear thermal expansion (CLTE) of the Nylon 66 at different temperature intervals. 25% to 30% decrease in CLTE can be achieved in Nylon 66 with the addition of HGMs up to 30 vol%. Likewise, Linear Mold Shrinkage (LMS) decreases as the amount of HGM loading is increased. Figure 3.22 shows a 13% reduction for injection molded Nylon 66 and a 50% reduction in LMS is observed in injection molded PP containing 30% by volume of 3M™ HGM iM30K.

Comparative Review of other Thermoplastic Weight Reduction Methodologies and Combinations with HGMs

There are several methods to reduce the density of an injection molded part (Figure 3.23). These include chemical foaming, supercritical foaming (e.g., MuCell®), adding lightweight fillers, thin walling, making composite structures, and material substitution.

Table 3.27 Comparison of Polyamide 66 (PA66) + 20 wt% Glass Fiber (GF) to Hollow Glass Microsphere Containing Formulations; GF Level Maintained

	Formula 1		Formula 2		Formula 3	
	PA66 Base Resin + 20 wt% GF		PA66 Base Resin + 20 wt% GF + 5 wt% iM30K		PA66 Base Resin + 20 wt% GF + 3 wt% iM16K	
Component	**Wt%**	**Vol%**	**Wt%**	**Vol%**	**Wt%**	**Vol%**
PA66	80	90	80	83.9	85.3	83.9
GF	20	10	20	9.8	20	9.7
HGM-iM30K	-	-	5.0	6.3	-	-
HGM-iM16K	-	-	-	-	3	6.4
GB/GF vol ratio	-		0.65		0.66	
Final %	100	100	100	100	100	100
Vol% glass	10		16.1		16.1	
Density (g/cc)	1.28		1.24		1.23	
Tensile strength (MPa)	138.5		149.2		151.7	
Tensile elongation (%)	5.4		6.1		6.2	
Tensile modulus (MPa)	3311		3502		3444	
Flexural modulus (MPa)	5453		6022		5811	
Room temperature Izod impact strength (kJ/m^2)	6.0		6.0		6.3	

Table 3.28 Comparison of Polyamide 66 (PA66) + 30 wt% Glass Fiber (GF) to Hollow Glass Microsphere Containing Formulations; Glass Fiber Level Maintained

	Formula 1		Formula 2		Formula 3	
	PA66 Base Resin + 30 wt% GF		PA66 Base Resin + 30 wt% GF + 5 wt% iM16K		PA66 Base Resin + 30 wt% GF + 10 wt% iM16K	
Component	**Wt%**	**vol%**	**Wt%**	**vol%**	**Wt%**	**Vol%**
PA66	70	83.9	65	71.4	60	60.8
GF	30	16.1	30	14.7	30	13.6
HGM-iM30K						
HGM-iM16K			5	13.9	10	25.6
GB/GF vol ratio	-		0.94		1.9	
Final %	100	100	100	100	100	100
Vol% glass	16.1		28.6		39.2	
Density (g/cc)	1.37		1.25		1.15	
Tensile strength (MPa)	191.4		176.9		164.1	
Tensile elongation (%)	6.6		6.3		5.4	
Tensile modulus (MPa)	4454		4434		4595	
Flexural modulus (MPa)	7470		7598		8044	
Room temperature Izod impact strength (kJ/m^2)	8.9		8.6		7.7	

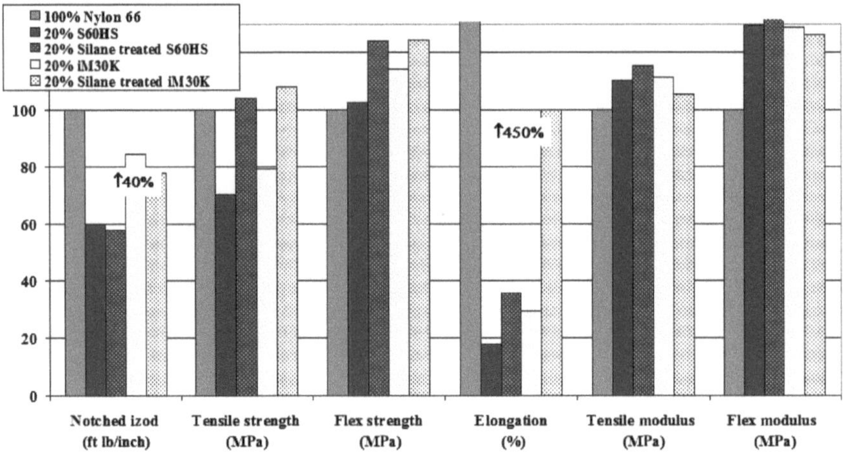

Figure 3.20 Physical properties for hollow glass microsphere coated with PA66 specific silane coupling agent.

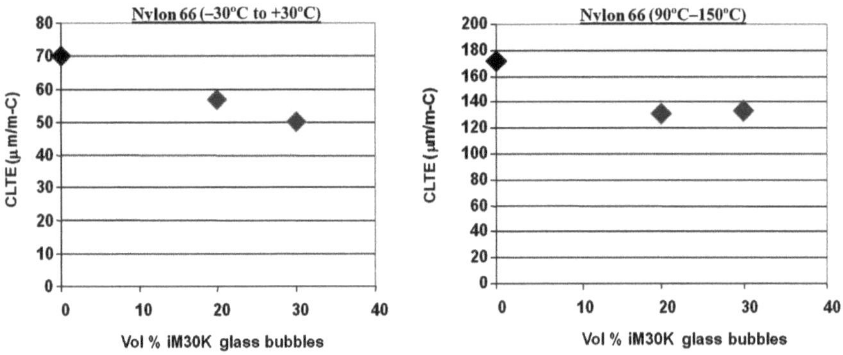

Figure 3.21 Coefficient of linear thermal expansion (CLTE) of nylon 66 as a function of hollow glass microsphere loading.

Thin walling (wall stock reduction) requires molding objects with wall thicknesses between 2.0 mm and 1.2 mm. The most challenging are those with wall thicknesses below 1.2 mm. Obvious benefits are (1) reduction in overall component size and weight, (2) reducing costs by reducing material use, and (3) faster cooling times and processing cycle times.

Considerations for thin walling are faster heat losses resulting in higher injection filling pressures. Wall thickness is always dependent on flow length, and the thickness reduction is limited by maximum injection pressure, the legal regulations, and the need for higher component stiffness.

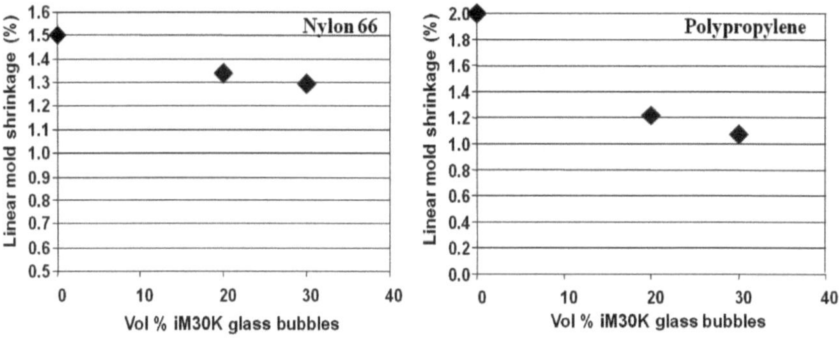

Figure 3.22 Effect of hollow glass microsphere loading on linear mold shrinkage.

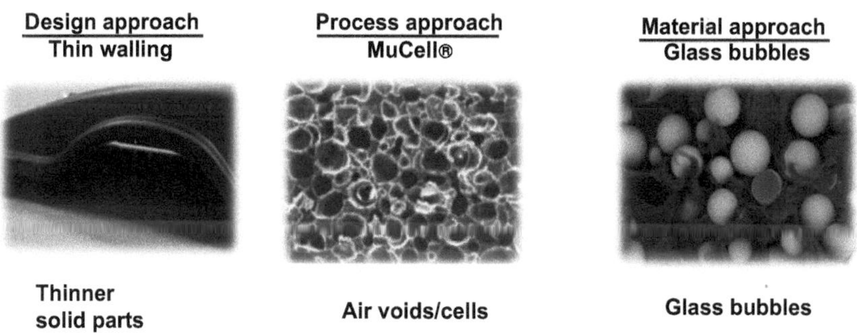

Figure 3.23 Weight reduction methodologies.

One can see in equation (3.3) that stiffness is a cubic function of thickness which indicates that stiffness reduces rapidly with decreasing thickness. A 20% decrease in thickness would decrease stiffness by about half.

$$\begin{aligned} \textit{Stiffness} &= (\textit{flex modulus}) \times \textit{moment of inertia} \\ \textit{Stiffness} &= E \times I = E \times \left(B \times \frac{t^3}{12}\right) \end{aligned} \tag{3.3}$$

where

E = flexural modulus (psi)

I = moment of inertia (in.)

B = element width (in.)

t = thickness (in.)

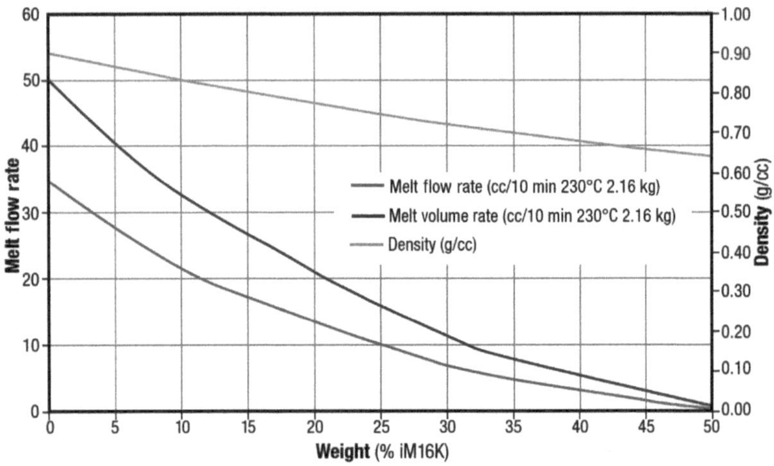

Figure 3.24 Melt flow rate of a high flow polypropylene as a function of wt% loading of 3M™ iM16K hollow glass microspheres.

Thin walling also requires resin viscosity modification such as peroxide "visbreaking" in PP to enable high flow rates into the mold cavity. However, it is also well known that visbreaking reduces molecular weight which has a negative effect on mechanical properties.

Figure 3.24 shows the MFR as a function of HGM (0.46 g/cc) loading in a high glow PP (55g/10 min @230 °C 2.1 kg). The reduction in MFR with HGMs should be taken into account when designing materials containing HGMs for thin wall injection molding applications.

Another method to reduce the weight of a molded part is achieved by reducing density of the parts via foaming. There are various foaming technologies but the foaming technology which is most well-suited for use with HGMs is microcellular supercritical foaming. This is because the particle size of the HGMs and supercritically generated foam cells are both in the micron range and in the same order of magnitude in size as seen in Figure 3.25.

The microcellular foam (MuCell®) injection molding technology involves the controlled use of gas in its supercritical state to create a foamed and hence a lightweight part with a lesser amount of polymer resin. The cells are created or nucleated as a result of homogeneous nucleation that occurs when a single-phase solution of polymer and gas in supercritical fluid (SCF) state (commonly nitrogen, but occasionally carbon dioxide) passes through the injection gate into the mold. The SCF is injected into the barrel where it is mixed with the polymer as shown in Figure 3.26. A shut off nozzle maintains the single-phase solution while the injection molding screw maintains sufficient back pressure at all times

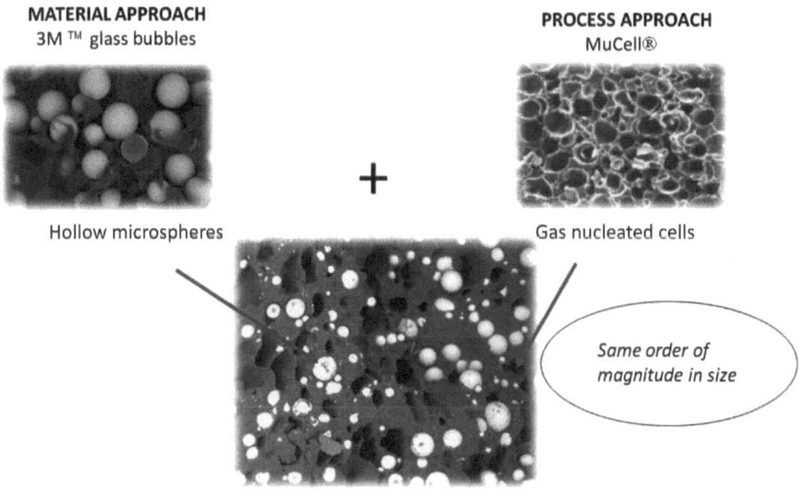

Figure 3.25 Scanning electron microscope images of hollow glass microspheres (upper left), gas nucleated cells (upper right), and combination of the two technologies (down) in plastics showing cell sizes that have similar order of magnitude in size.

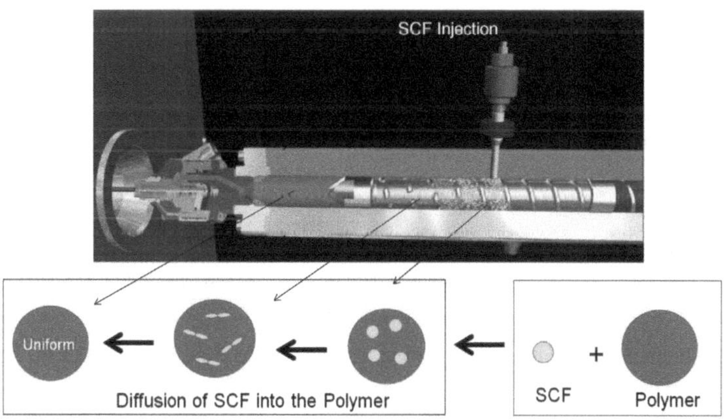

Figure 3.26 MuCell® process-supercritical gas injection. *Courtesy of Trexel Inc.*

to prevent premature foaming or the loss of pressure which would allow the single-phase solution to return to the two phase solution.

Today injection molding machines can readily be provided with, or retrofitted for microcellular supercritical foaming capability. Figure 3.27 shows an injection molding machine with Mucell® foaming components comprising (1) SCF injector, (2) nonreturn valve, (3) SCF interface kit, (4) SCF metering system, and (5) gas supply.

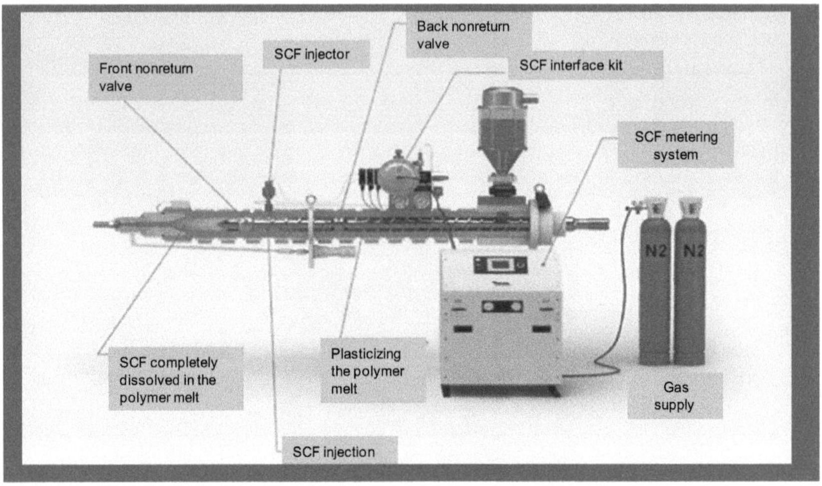

Figure 3.27 Injection molding machine with Mucell® foaming components. SCF stands for supercritical foaming. *Courtesy of Trexel Inc.*

Most obvious benefits of molding with supercritical foaming over solid (nonfoamed) molding is the use of a smaller shot size since the foam cell growth enables final mold fill. In addition, hold time or pressures are highly minimized with results in reduced molded-in stress, less warp and bow, lower clamp tonnage, and consequently less tool wear. There is also no need to adjust runner/gates size for pack pressure. Considerations for weight reduction technology via SCF are the initial capital cost versus return on investment and, in some cases, cosmetic issues.

There are several benefits of HGMs when combined with SCF molding. Small amounts of HGMs improve cell structure (more frequent, uniform, and smaller) increasing the efficiency of a given SCF. HGMs add further weight reduction not only because they are low-density fillers but also due to improved cell nucleation. With improved cell nucleation in the presence of HGMs, part filling via expansion of the growing gas bubble is more efficient allowing one to set a smaller shot size with HGMs. HGMs improve warpage because they are isotropic while SCF molding improves warpage due to little or no hold pressure used. The end product with the combination of technologies has great dimensional stability. HGMs reduce cooling time due to reduced volumetric heat capacity of the part (less resin to cool) and SCF molding reduces cooling time with the elimination of most of the solid pack hold time. Also, higher temperatures, typically needed for amorphous polymers to minimize residual stresses caused by the decreasing gradient of pack pressures from gate to end of fill, are not needed with SCF molding. HGMs also cause scattering of the light and bring about a more uniform color appearance in white to light colored SCF foamed parts.

When HGMs are used in combination with SCF molding, it is possible to achieve double digit density reductions with well-maintained properties. This is true especially for reinforced thermoplastics (such as GF) where formulations can be optimized by reformulating the amount of reinforcing filler and HGMs for best property retention across all modes of deformation, that is, tensile, flex, and impact.

This is demonstrated below by a case study targeting such applications as automotive engine covers and other under the hood components utilizing GF filled nylon 6. The description of the materials employed in this study is shown in Tables 3.29, 3.30, and 3.31.

Test specimens were molded in a Mucell® enabled Engel injection molding machine ES200/100 TL equipped with SCF system Model # TR

Table 3.29 Unfilled Polyamide 6 (PA6) Provided by BASF

Product name	**Ultramid® 8202HS BK 102**
Description	Unfilled, heat stabilized, injection moldable, black PA6
Density	1.13 g/cc

Table 3.30 Polyamide 6 (PA6)-25HGM Masterbatch (3M)

Product Name	**3M iM16K Masterbatch**
Description	25% Hollow glass microsphere filled ultramid 8202HS BK base, gray PA6 resin
Density	0.88 g/cc

Table 3.31 Polyamide 6 (PA6)-30 wt% GF Provided by BASF

Product Name	**Ultramid B3WG6 GIT BK 807**
Product description	30% Glass filled, heat stabilized, injection moldable, black PA6
Density	1.36 g/cc

3.5000G. Dry blend of pellets at the injection molding step was used to prepare blend ratios desired. Tables 3.32 and 3.33 show the ratio of the compounds used in various dry blend ratios to prepare PA6-15 GF and PA6-18 GF-8 HGM, respectively (Figure 3.28).

Table 3.32 Dry Blend to Achieve PA6-15 GF

	Density	Weight	Volume
Component	**g/cc**	**%**	**%**
PA6-30 wt% GF	1.3558	50.000	45.458
Unfilled PA6	1.1300	50.00	54.54
Final compound	1.2326	100.000	100.000

Table 3.33 Dry Blend to Achieve PA6-18 GF-8 HGM

	Density	Weight	Volume
Component	**g/cc**	**%**	**%**
PA6-30 wt% GF	1.3558	61.000	51.096
Unfilled PA6	1.13	5.00	5.03
PA6-25 HGM	0.88	34.00	43.88
Final compound	1.1357	100.000	100.000

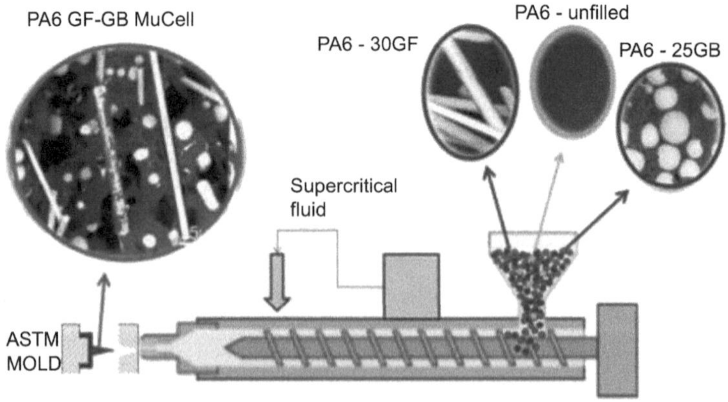

Figure 3.28 Dry blend of pellets at the injection molding step.

Table 3.34 Mechanical Properties of Glass Fiber (GF) Reinforced Heat Stabilized Nylon 6 in the Presence of Hollow Glass Microsphere (HGM) and Supercritical Fluid (SCF) MuCell Foaming

	Formula 1	Formula 2	Formula 3	Formula 4
	Wt%	Wt%	Wt%	Wt%
PA6	85	74	74	74
GF	15	18	18	18
iM16K-HGM (0.46 g/cc)	-	8	8	8
MuCell vol%	Solid	Solid	6.4	8
Density (g/cc)	1.230	1.140	1.062	1.050
Tensile strength (MPa)	113	124	99	93.3
Tensile Modulus (MPa)	4150	5110	4480	4275
Flexural Modulus (MPa)	4117	5170	4620	4415
Flexural strength at yield (MPa)	165	185	150	145
Notched Izod impact (J/m)	40	46	54	53

Table 3.34 compares mechanical properties and density of 15% GF filled virgin nylon 6 compound (Formula 1) to 18 wt% GF and 8 wt% HGM compound with HGMs only (Formula 2), and HGM and MuCell combined with increasing foaming levels (Formulas 3 and 4). As shown in Formula 2, one can achieve 7.3% weight reduction with HGMs only, while improving all properties considerably. This is because increasing GF amount slightly from 15 wt% to 18 wt% helps improve the properties of the compound significantly but has little effect on increasing density and the density reduction is achieved by the HGMs which also add to the modulus.

Formulas 3 and 4 show that increasing MuCell physical foaming levels decreases the stiffness of the compounds but the combined MuCell and

HGM compounds still exhibit higher modulus (stiffness) than the control sample. For instance, Formula 4 shows 13.7% density reduction with well-maintained stiffness properties ±10%, that is, 8% increase in tensile modulus, 12% increase flex modulus, 9% decrease in flexural strength, 11% decrease in tensile strength. Impact strength in all HGM/MuCell compounds is increased (~30%) compared to the control and HGM only formula which is due to the foaming effect. Formula 4 presents a 14.6% weight reduction with well-maintained properties (Figure 3.29).

Another case study for the combination of HGMs with SCF molding (MuCell®) is shown in Table 3.35 for chemically coupled heat stabilized PP targeting under the hood applications such as fan shrouds. About ~15% density reduction is achieved in Formula 2 in comparison with Formula 1 while improving tensile and flexural modulus. Modulus levels are well maintained even with 17% density reduction achieved via HGM and SCF combination in Formula 3 while the modulus in Formula 4, utilizing no HGM, no SCF, and reduced GF amount, is dramatically reduced. Flexural strength is also well maintained (±10%) in these parts while the impact strength is reduced in HGM containing formulas (20–25%) but it is still well within the range used for these low impact stiff parts.

Table 3.36 summarizes the individual and combined HGM and SCF benefits in GF filled PP and PA and shows methodologies to optimize properties. Red, yellow, and green color indicates a negative, neutral, and positive attribute while the equal, plus, and negative sign indicates no

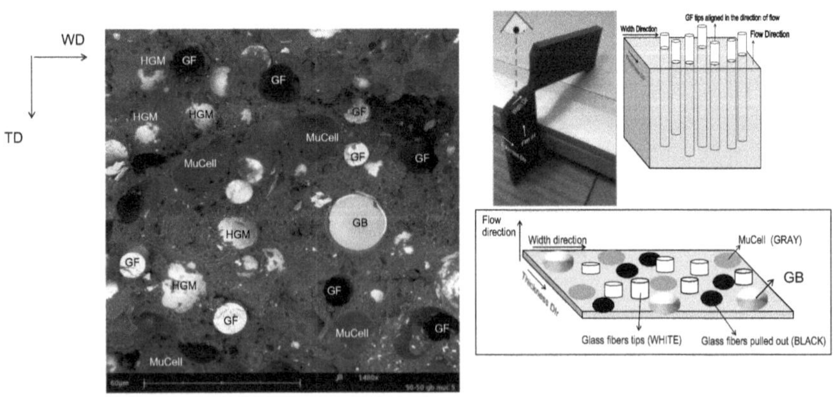

Figure 3.29 Scanning electron microscopy of a impact bar cross section of a nylon/glass fiber (GF)/hollow glass microsphere (HGM) compound with MuCell.

Table 3.35 Mechanical Properties of Glass Fiber (GF) Reinforced Heat Stabilized Nylon 6 in the Presence of Hollow Glass Microsphere (HGM) and Supercritical Fluid (SCF) MuCell Foaming

Component		1-Control		2		3		4	
		PP 30 GF		PP 31.5 GF/9 HGM		PP 29.5 GF/10.5 HGM		PP 15 GF	
		Wt%	Vol%	Wt%	Vol%	Wt%	Vol%	Wt%	Vol%
Polypropylene (PP)		70	86.8	59.5	63.3	60	61.9	85	94.1
GF		30	13.2	31.5	11.9	29.5	10.8	15	5.9
Surface treated HGM iM16K (0.46 g/cc)		-	-	9	16.9	10.5	18.8	-	-
SCF cell occupied vol%		Solid		7.3		8.5		Solid	
Density (g/cc)		1.122		0.957		0.930		1.000	
% Reduction		0.0		−14.7		−17.1		−10.9	
Tensile modulus (MPa) D638	Specific modulus (MPa/g/cc)	4033	3594	4400	4598	4120	4430	2715	2715
Flexural modulus *(MPa) D790 (tangent)	Specific modulus (MPa/g/cc)	4890	4358	5334	5574	5300	5699	3040	3040
Flexural strength (MPa) D790		107.9		98.5		94.0		80.0	
Room temperature Izod impact strength (J/m) D256		100		74		80		77	

Table 3.36 Individual and Combined HGM and SCF Benefits in Glass Fiber Reinforced PA and PP

	PA or Homo PP Base Resin	GF	GF+HGM	GF + SCF	GF+HGM+ SCF	1.2GF+ HGM+ SCF
Density	=	+	=	=	-	-
Tensile Strength	=	++	+	+	+	++
Tensile Modulus	=	++	+++	+	++	++
Flexural Strength	=	++	+	+	+	++
Flexural Modulus	=	++	+++	+	++	++
Notched Impact	=	++	+	+++	++	++
Warpage	=	++	+	-	-	-
Mold Cycle Time	=	=	-	-	--	--

GF: Glass Fiber HGM: Hollow Glass Microsphere SCF: Supercritical Fluid PA: Polyamide PP: Polypropylene

significant change, increase, and decrease in that attribute. Multiple plus or minus signs indicate significantly increased and decreased attributes. Note that this table indicates a trend more than the magnitude of that trend. For instance, in some cases the increase or decrease of modulus may not be very significant in GF filled systems either with SCF or HGM.

In the preceding sections, we have discussed HGM use and benefits in two major polymer systems, that is, PP and nylon. In addition to these polymer systems, HGMs can be used for a variety of reasons in more application specific examples. One of these areas is in polymer wood composites (PWC). Processing behavior, morphology, and benefits of using low-density HGMs in PWC, that is, HDPE/wood flour, were discussed earlier [9]. Certene HHGM-0760 HDPE homopolymer in the form of pellets (Channel Prime Alliance) was the polymer used in the composites. STRUKTOL® TPW 104, a blend of lubricants designed specifically for wood fiber/flour filled polyolefins was used as the lubricant. Wood flour was a mixture of several wood types (pine, maple, and so on.) and 3M™ HGM iM30K and K42HS were used in extruded profiles. The profiles were prepared in a two-step process depicted in Figure 3.30. First, a pre-compound formulation without any HGMs was prepared in a TSE. The wood flour and lubricant was dry blended and side stuffed, while the HDPE was added in the main feed throat. The strands were pelletized into pre-compound-pellets. In the second stage, the Precompound-pellets were introduced in the main feed throat, while the HGMs were side stuffed into the extruder using the following formulations. The screw speed of the extruder was set to 50 rpm. The zone temperatures ranged from 360 F to

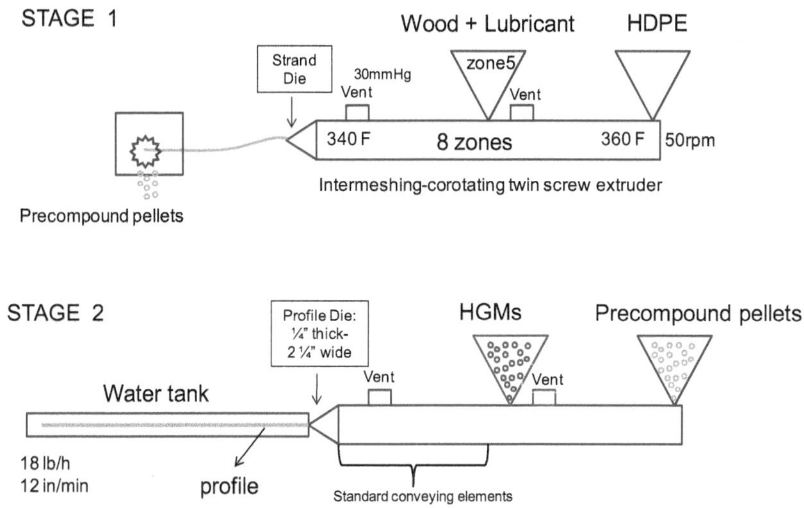

Figure 3.30 Hollow glass microsphere (HGM) two-step compounding process in high density polyethylene (HDPE) wood composites.

340 F. Profiles were extruded at 12 in/min (18 lbs/h) through a 2¼ in × ¼ in die, water cooled, and dried at ambient temperature.

Table 3.37 shows the HGM formulation used in the HDPE wood composites.

In this study, the density of HDPE/wood profiles was successfully reduced to below 1.0 g/cc allowing them to float in water (Figure 3.31).

SEM images evidenced uniform dispersion and high survival rates of HGMs in PWC (Figure 3.32).

The presence of HGMs increased the flexural modulus while decreasing density (Figure 3.33). Impact properties are preserved at 70% of the original value. HGM filled formulations also showed lower thermal conductivity (Figure 3.33).

One of the drawbacks of standard PWC is that it is harder to nail and screw into the composite material as compared to regular wood. This is mainly due to the presence of the consolidated polymer phase versus a void volume containing cellulose phase in regular wood. With the incorporation of HGMs into the wood composite, the composite accepts screws and nails more like real wood than do their counterparts without HGMs. In order to demonstrate this effect, a simple experiment was performed. Two profiles, one with 15 wt% HGMs and the other without any HGMs, were drilled using a heavy duty drill with a ¼" drill bit. The electric drill was held on the wood composite profiles without applying

Table 3.37 Formulations Used in Preparing Hollow Glass Microsphere (HGM) Filled High Density Polyethylene (HDPE)/Wood Composites

Formulation # - Precompund (GB type)	WF content (%)	Lub content (%)	HDPE content (%)	Glass Bubble content (%)
1-A (IM30K)	50.0	6.0	44.0	0.0
2-A (IM30K)	47.5	5.7	41.8	5.0
3-A (IM30K)	45.0	5.4	39.6	10.0
4-A (IM30K)	42.5	5.1	37.4	15.0
5-A (IM30K)	40.7	4.9	35.9	18.5
6-A (K42HS)	45.0	5.4	39.6	10.0
Precompound	Wood content (%)	Lub content (%)	HDPE content (%)	Bubble content (%)
A	50	6	44	0

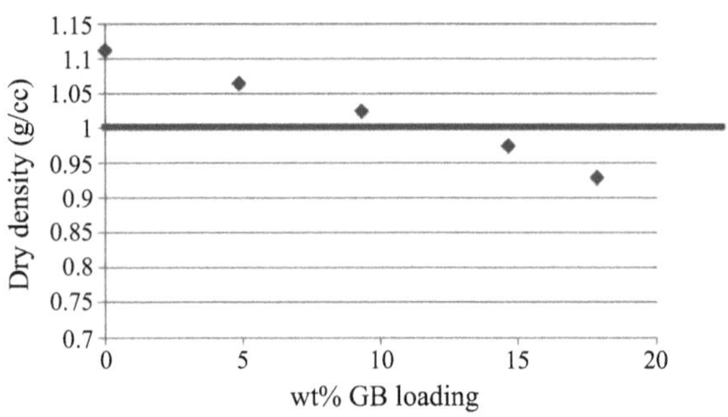

Figure 3.31 Density of high density polyethylene (HDPE)/wood composites as a function of HGM loading.

any extra pushing force other than the weight of the drill. One can approximate the pressure at the wood-drill bit contact point by dividing the weight of the drill by the cross-sectional area of the ¼" drill bit. 1800 grams drill weight applies a pressure of ~80 psi at the contact point. With this pressure, it takes about 1 min and 15 s for the drill to go through the ¼" thick profile without any HGMs while it is only 15–20 s for the profile with the HGMs. The HGMs, when broken, due to contact with the drill bit,

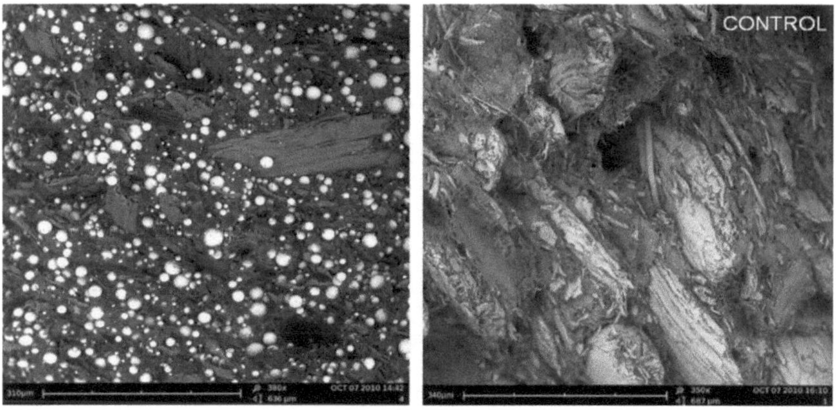

Figure 3.32 Scanning electron microscopy of hollow glass microspheres (HGMs) in wood plastic composite (HGM filled–left, Control–right).

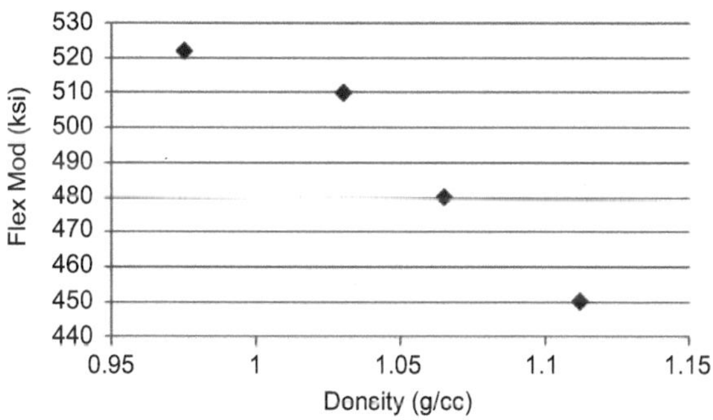

Figure 3.33 Flexural modulus as a function of density.

afford void volume for the drill bit to penetrate through. The same observation is true with the nails. Nails get inserted with much less effort into the composites with HGMs. HGMs also render sharper contours and corners in the extruded profiles as well as better surface definition than profiles without any HGMs. The addition of large amount HGMs causes the color of the wood to turn lighter (Figure 3.34).

From a processing point of view, HGMs do not increase the melt viscosity despite increased filler loading. Extruder load decreases with increased HGM content. Increased cooling rates are achieved in HGM filled composite profiles that could result in faster production rates.

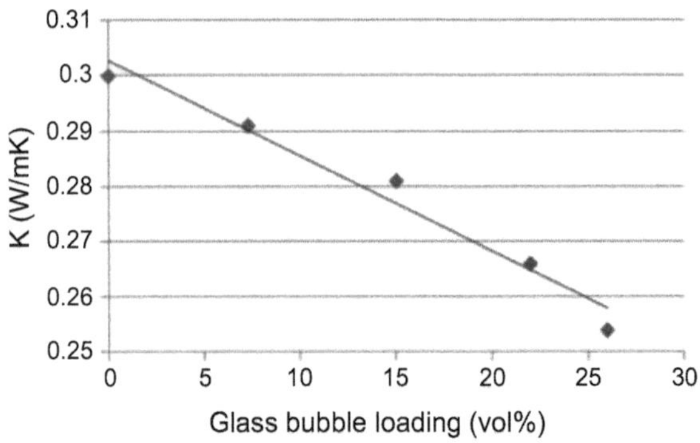

Figure 3.34 Thermal conductivity as a function of hollow glass microsphere loading.

References

[1] Section Entitled "Remove Excess Weight". http://www.fueleconomy.gov/feg/driveHabits.shtml.

[2] 3M Case Study "Glass Bubbles Reduce Weight of TPO Parts" 3M Website.

[3] 3M Case Study "Flying High with 3M™ Glass Bubbles" 3M Website.

[4] A. Salazar, On thermal diffusivity, Eur. J. Phys. 24 (2003) 351–358.

[5] G. Yang, A.D. Migone, K.W. Johnson, Heat capacity and thermal diffusivity of a glass sample, Phys. Rev. B 45 (1992) 157–160.

[6] T.P. Melia, The specific heats of linear high polymers, J. Appl. Chem. 14 (11) (November 14, 1964) 461–478, http://dx.doi.org/10.1002/jctb.v14:11.n/issuetoc.

[7] 3M Technical Paper – Productivity Benefits of 3M™ Glass Bubbles in Injection Molded Thermoplastics via Increased Cooling. http://solutions.3m.com/wps/portal/3M/en_US/Glass/Bubbles/Resources/Literature/.

[8] D.O. Kazmer, Injection Mold Design Engineering, Carl Hanser Verlag GmbH & Company KG, November 12, 2012.

[9] B. Yalcin et al, Journal of Plastic Film & Sheeting 28(2) (January, 2012) 165–180.

[10] J. Fischer, Handbook of Molded Part Shrinkage and Warpage, Elsevier, December 2012, ISBN 978-1-4557-2597-7.

[11] 3M Technical Paper "3M™ Glass Bubbles iM16K for Reinforced Thermoplastics". http://solutions.3m.com/wps/portal/3M/en_US/Glass/Bubbles/Resources/Literature/.

[12] 3M Technical Paper "Effect of Processing Conditions on the Extent of Glass Bubble Survival during Twin Screw Compounding". http://solutions.3m.com/wps/portal/3M/en_US/Glass/Bubbles/Resources/Literature/.
[13] U. YIlmazer, et al., Effects of processing conditions on the fiber length distribution and mechanical properties of glass fiber reinforced nylon-6, Polymer Composites 23 (1) (February 2002).
[14] 3M Technical Paper "Plug-n-play Weight Reduction Solution by Hollow Glass Microspheres Technical Paper". http://solutions.3m.com/wps/portal/3M/en_US/Glass/Bubbles/Resources/Literature/.
[15] P. Steve Chum, et al., Olefin polymer technologies-History and recent progress at the Dow Chemical Company, Progress in Polymer Science 33 (2008) 797–819.

4 Hollow Glass Microspheres in Rubbers and Elastomers

Stephen E. Amos

Hollow glass microspheres (HGMs) are used in many elastomeric applications—from shoe soles and tires to hoses and wire and cable compounds, from thermoplastic elastomers (TPEs) to liquid silicone rubber (LSR) sealants and void fillers. Often the main benefit is weight reduction, especially important for transportation applications. Insulation, stiffening, and cycle time reductions are additional attributes afforded by HGM for transportation and other applications. In general, they are used for many of the same reasons as discussed in the Thermoplastics chapter, but physical property changes are somewhat different.

Rubber is basically a thermoset elastomer, using a curing or cross-linking mechanism to generate high molecular weight which in turn creates the desired mechanical properties. There are many, chemically diverse rubbers including hydrocarbon, nitrile, fluorinated, chlorinated, silicone, and ether moieties. One property characteristic of all is elasticity, or the ability to be stretched and deformed without yielding and eventually returning to its original shape. Historically the oldest rubber—natural rubber—was produced by "tapping latex" from the rubber tree (*Hevea brasiliensis*), a natural source of monomer *cis*-polyisoprene (Figure 4.1).

Rubber compounds are viscoelastic materials meaning they have both a viscous and elastic component to their behavior. The elastic component is like a spring, which when compressed and uncompressed, exerts and recovers close to the same amount of energy. The viscous component provides energy dissipative behavior by converting the loss in recovered kinetic energy into heat. Depending on the type of rubber polymer selected, as well as the compounding ingredients, one can create

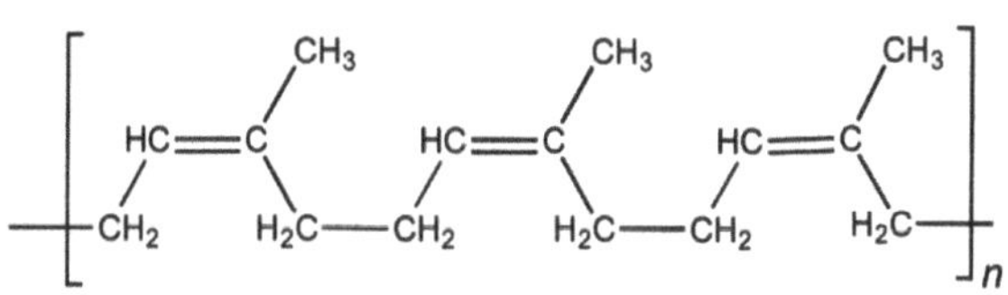

Figure 4.1 *Cis*-polyisoprene.

Hollow Glass Microspheres for Plastics, Elastomers, and Adhesives Compounds
http://dx.doi.org/10.1016/B978-1-4557-7443-2.00004-9

compounds that cover a range of responses, from mostly elastic to mostly viscous. Elastic materials provide a faster more complete energy recovery response, like a "superball," which when dropped, bounces back to near the height from which it was dropped. A viscous compound by comparison, although flexible like rubber, might only bounce back to 10% of its original drop height. Although these kinds of materials do not provide good energy recovery, they have the advantage of being able to dissipate energy which can help minimize peak stresses at crack locations, and result in very tough materials that do not easily tear.

Benefits of HGM Use in Rubber

The main benefit of using HGMs in rubber is density modification—generally reducing density or weight of automotive seals, gaskets, or molded parts. This is becoming very important in transportation applications for fuel economy and CO_2 emission reduction. But density modification is also important for buoyancy in water for floating fuel hose, naval bumpers, and floating golf balls to name a few known examples. It can also be used for creating buoyancy in other liquids such as gasoline for fuel tank floats or bladders. Being able to "dial in" a density to a compound or a specific component of a composite also plays a role in diverse applications such as differential weighting in golf balls, fishing lures, and recyclability. Often in sporting equipment applications differential weighting provides for improved motion or further distance attainable with the articles. For recycling, ability to tune a density can be used in density gradient columns to separate mixed resin streams.

A secondary benefit for rubber systems is changing the insulation properties of the composite article. Some rubber materials such as LSR are used in deep water oil and gas production for pipe, tree, and remote operated vehicle thermal insulation applications. Use of HGMs in these systems can reduce thermal conductivity (K value) of the composite 30–40% from the unmodified base resin. Sound dampening and acoustical insulation (and light weighting) properties are provided for rubber trunk mats and seam sealers in cars, trucks, and airplanes. Finally, improved electrical insulation and dielectric properties are possible in TPE and other rubber based wire and cable compounds.

A third benefit in rubber processing is the reduction in thermal mass of the composite allowing for faster cooling cycles. By adding HGMs to rubber compounds a large amount of polymeric material can be removed, replaced by largely void space. The amount of heat energy required to make a set volume of the uncured material flow through an extrusion process, for example, is less so the subsequent downstream operations—cooling, chopping, coiling, etc., can be done sooner, allowing for increased line speeds and productivity.

Physical Property Changes

Physical properties of the cured rubber composite will change with increasing level or parts per hundred (PHR) of HGMs. Some are considered a benefit for a given application. An example of this is stiffness or modulus increase in tire tread compounds. Adding stiff glass particles to a very ductile rubbery system can increase the stiffness of the final article. Adding stiffness to tire treads can improve run flat performance and reduce rolling resistance, improving fuel economy. Though not used for tire treads a demonstration of tensile modulus increase with high loading of HGM in fluoroelastomer is shown in Figure 4.2. This fluoroelastomer system was loaded with increasing amounts of 0.46 g/cc, 114 MPa (3M HGM iM16K a high crush strength bubble designed for plastic and rubber applications) and the same system will be used to demonstrate other physical property changes. Measurements were made after two cure conditions—a press cure and a post cure to generate higher cross-link density.

Other physical property changes may be less desirable such as a shorter elongation to break. At low loadings (<15 volume %), in some elastomeric systems, the elongation at break is not negatively affected. At high

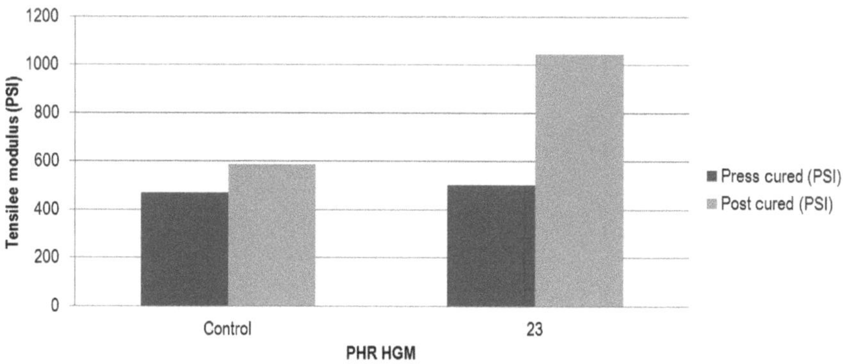

Figure 4.2 Tensile modulus of 42 mooney fluoroelastomer with increasing loading of HGM, press cured for 10 min at 177 °C, post cured for 16 h at 232 °C.

volume loadings the elongation might reduce which is a general rule with any filler material added at high volume percent. In the same fluoroelastomer system, Figure 4.3 shows the response of elongation to break with increasing levels of HGMs. Note the drop in elongation percent for the 23 PHR of the HGM sample.

Compression set is a critical property for sealing applications of rubber. Compression set measures the failure of a rubber material to return to its original dimensions after a compressive load has been applied for a period of time, usually at elevated temperature. If HGMs break under the compressive load this could lead to volume loss and an increase in the measured compression set. Knowing the compressive force applied on the seal, selection of the correct strength HGM (higher crush strength than the applied force) and good dispersion of the HGM is critical for compression set performance. Recent testing of a 0.46 g/cc specific density, 114 MPa HGM in a 42 Mooney viscosity fluoroelastomer showed good compression set performance at various PHR loadings when tested at 25% deflection for 22 h at 200 °C (Figure 4.4). Given the error in this test method these compression sets are essentially the same.

Rubber is often used for applications requiring wear resistance. Wear resistance is commonly measured by weight or volume loss upon abrasion with an abrasive wheel or other device. Three common methods are Akron, NBS, and Tabor abrasion testing. Since HGMs typically add stiffness to a rubber compound they generally cause the abrasion resistance to drop as discussed in the Shoe Sole Application Example below.

Tensile strength is also an important physical property to consider in a filled rubber composition. A similar response to elongation is seen for

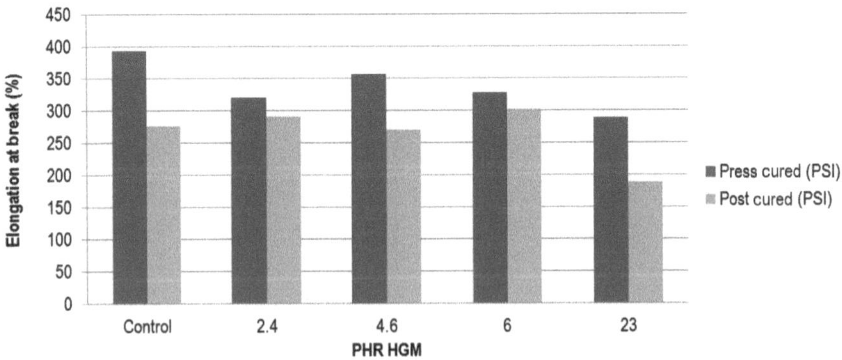

Figure 4.3 Percent elongation at break of a 42 mooney fluoroelastomer with increasing loading of HGM, press cured for 10 min at 177 °C, post cured for 16 h at 232 °C.

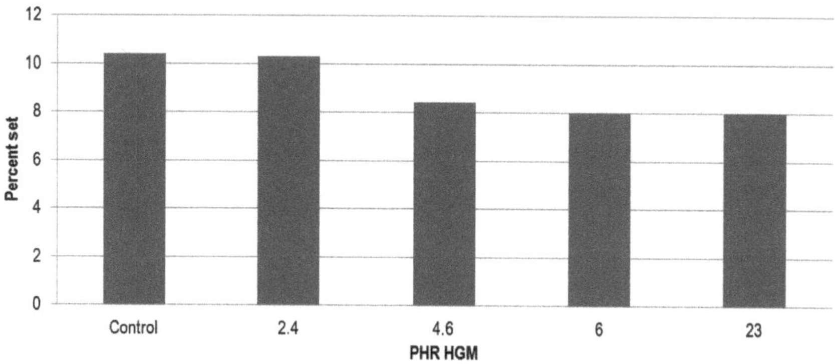

Figure 4.4 Compression set of a 42 mooney fluoroelastomer with increasing loading of HGM, press cured for 10 min at 177 °C, post cured for 16 h at 232 °C.

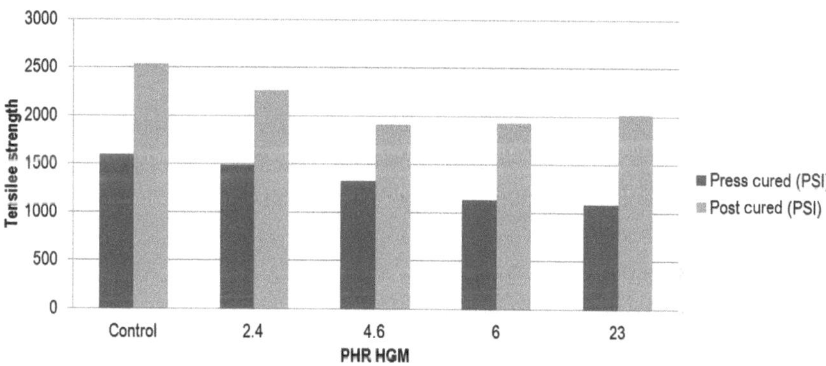

Figure 4.5 Tensile strength of a 42 mooney fluoroelastomer with increasing loading of HGM, press cured for 10 min at 177 °C, post cured 16 h at 232 °C.

tensile strength – at low loadings, little change, at high volume percents the strength can reduce significantly as shown in Figure 4.5.

Incorporation of HGMs in Rubber

As noted before, if the HGMs break, the compound density increases because the void volume is lost and resulting pieces of glass from the broken sphere have a significantly higher density (for soda lime borosilicate glass this density is 2.54 g/cc). With rubber compounds containing many ingredients that either are inorganic in nature or contain metal ions that oxidize and become nonvolatile at ash temperatures; calculating bubble breakage from ash density can be difficult and have many sources of error. If the HGMs are incorporated into the host rubber with minimal breakage, downstream processes have significantly less potential to cause additional breakage.

As with all high shear, high pressure processes, the key to effectively adding the HGMs is to do it in a way that does not break them causing volume loss. There are several ways HGMs may be damaged during incorporation in to rubber. Addition of the HGMs to the mixing process is one variable. If the neat HGMs are added together with other hard non-melting, hard materials (fillers, fibers, etc.,) they might be broken due to particle on particle impact or stress as they are sheared and mixed together. If the neat HGMs are added to the resin but not dispersed well they may still encounter hard particles as the process applies more pressure. Imagine several hollow eggs in a box being shoved around and dropped during shipping vs. those same eggs, in JELL-O®, being handled similarly. The JELL-O is analogous to the resin and provides two benefits; (1) cushioning sudden impacts or momentum change and (2) providing interstitial material preventing egg-on-egg contact. This latter point is even more profound when you consider the spherical shape. Sphere-on-sphere contact is a tangent resulting in little area over which to spread the force.

Typically rubbers are compounded with curatives and many additives and fillers via one or more of three main processes: (1) milling, (2) mixing (external types like kneaders and internal types such as a Banbury® mixer) and (3) extruding. Selection of the compounding process is usually made by considering the amount of distributive or dispersive mixing required. There is also a requirement for the compounding to mesh with downstream processes. For example, a hose manufacturing process may use a Banbury to compound the rubber, and then it is transferred to an extruder so that the compound can be shaped via a die into a hose, then finally heat treated by passing through ovens to cure.

2-Roll Mills

The method of incorporating HGM in to the rubber is one of the critical steps for maintaining the initial density of the sphere and the low density of the composite rubber compound. A 2-roll mill is a fairly straightforward way of understanding the potential for HGM breakage. Most grades

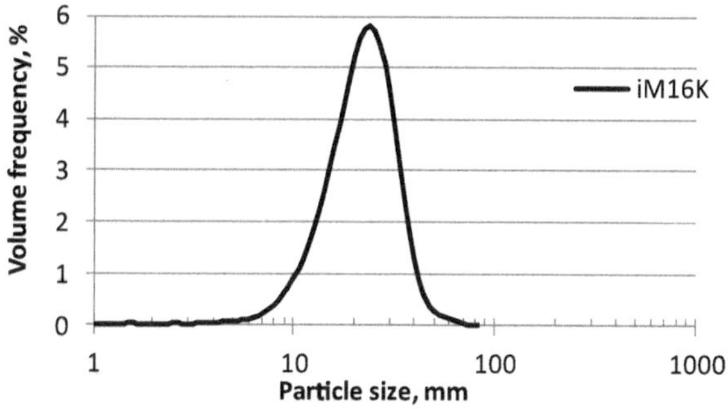

Figure 4.6 Particle size distribution for 3M™ hollow glass microsphere iM16K.

of HGMs have a range of particle size that usually is normally distributed. For example, the 3M™ Hollow glass microsphere iM16K, targeted for these types of applications has the particle size distribution shown in Figure 4.6.

If the top size (D99) particle size for this HGM is roughly 50 μm then the gap between the rolls in a 2-roll mill needs to be at least a 5−10x multiple of that dimension (0.5 mm or 10x as pictured in Figure 4.7) to ensure the HGM material is not broken during processing. There are other factors to consider as well as the roll gap, including viscosity/molecular weight of the rubber phase, volume % of other hard, nonmelting particles, shear rate of the mill, roll heating/temperature, addition sequence of additives, use of knives and skiving equipment, and chemical interactions

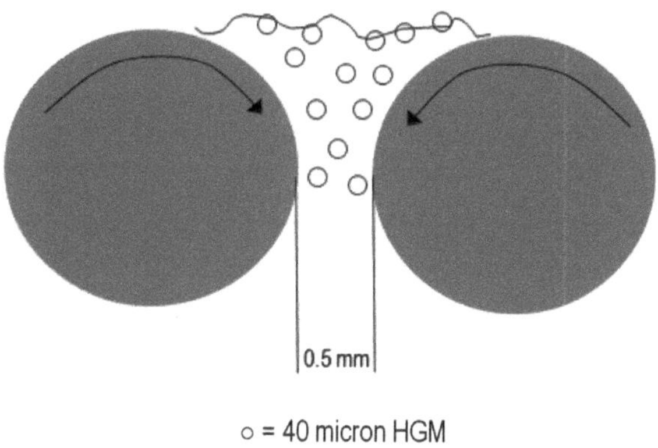

Figure 4.7 Roll gap in 2-roll mill processing allows for HGM processing with minimal breakage.

with other additives. Furthermore, due to the low density and high volume occupied by the HGMs it is often necessary to stage the addition on a 2-roll mill to efficiently fold them in to the rubber and prevent them from spinning on top of the rubber mass between the rolls.

Of the other factors listed probably the second major parameter to consider is the viscosity/molecular weight of the rubber phase. A recent study of different rubber materials with a constant volume % of an HGM that has a crush strength that can be on the edge of survival for this type of processing (52 MPa) shows significantly different volume loss with respect to molecular weight, viscosity, and sheer thinning behavior of the rubber phase (Table 4.1).

Internal Mixers

The use of internal mixers such as the common commercially used Banbury Mixer (schematic in 63)—has other considerations for HGM addition. On the one hand, a closed mixing setup is preferred for low density additives because it contains the material and prevents it from being dispersed around a factory and coming in contact with workers (most HGM products have no serious skin contact or hazardous health or

Table 4.1 Effect of Rubber Type and Mooney Viscosity on HGM Breakage

Rubber	Type	Mooney ML	HGM Volume Loss%
SVR-3L	NR	>80	74
NBR-25LM	NBR	50	33.4
NBR-40H	NBR	80	55.5
X-750	X-NBR	47	44
SBR-1502	SBR	52	11
E-30	SBR	NA	31
KEP-330	EPDM	28	Negligible
VCR-412	BR	45	Negligible
VCR-617	BR	62	Negligible
KBR-01	BR	45	Negligible

HGM, hollow glass microspheres.

inhalation issues but they can be a nuisance and inhalation or contact should be kept to a minimum by wearing proper PPE). On the other hand, closed mixers can hide poor distribution and dispersion issues and may allow build up on the internal low shear spaces of the mixer. One strategy for preventing this is to do an upside-down mix addition. This is where the additives and oils are added first and blended for a short time and then the polymer is added to sweep the void spaces of the mixer clean. This is typically used with high volume % filler loadings or high liquid levels, or both. Typically, even low concentration HGM containing formulations rise to this level due to the low density of the additive.

Selection of the rotor blade is also a consideration to reduce or prevent HGM breakage in internal mixers. There are four common blades to choose from pictured in Figure 4.8 with increasing shear mixing capability from left to right. Some HGM grades are recommended for rubber milling processes due to their high isostatic crush strength (e.g., 3M™ Hollow glass microsphere iM16K). Commonly used rubber mixing blade designs, such as cam and roller blades, can be used with these HGMs. Sometimes an unusually low density is required (e.g., making dense rubber like chloroprene or fluoroelastomer float). In these cases a lower density, lower strength bubbles may be used with a low shear design like the sigma blade.

Other Incorporation Methods

For high molecular weight rubbers two other incorporation methods are extrusion and kneading devices. Many of the same principles discussed in the Thermoplastics chapter about extrusion are applicable for the uncured

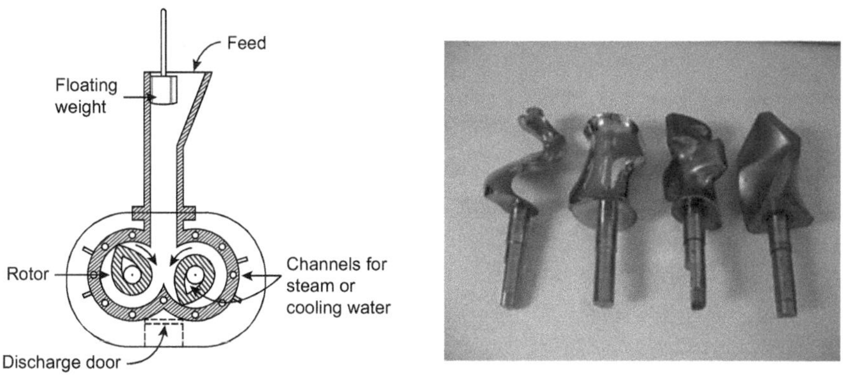

Figure 4.8 Banbury Mixer and common rotor designs (pictured from right to left increasing shear blades; sigma, banbury, cam and roller blades).

rubber system. The use of downstream feeding through side stuffing, modifying the screw to have low shear mixing and/or just conveying elements after HGM addition are recommended. Kneaders have similar blade designs but are not enclosed like Banbury mills. Here the shear issues are the same HGM selection but the correct PPE must be used because some bubbles will be lost to the surroundings.

For thermosetting rubber systems that start as low viscosity liquids the same methods of mixing discussed in the *Mixing chapter* are relevant. The intensity of the mixing blade or pump will dictate the density and strength of the HGM that can be used.

Rubber Additive Formulations and HGM Considerations

Rubber compound formulations have generally evolved over many years with new additives and fillers combined with old. A typical rubber formulation may have two or three different types of rubber adding up to 100 parts and then curatives, fillers, process stabilizers, accelerators, antiozonants, antioxidants, lubricants, processing aids, acid and base neutralizers, and colorants, to name a few, are added on top in the old part per 100 ("PHR" of 100 parts of rubber) formulation scheme. Understanding all of the potential interactions and imparted attributes of these additives, during processing, curing and in the end-use application, is more of an art than a science. At the very least it would require a large design of experiment to map the interactions of a multicomponent system such as this and so much of this knowledge is learned by experience and maintained cerebrally. To show the complexity, Table 4.2 is an example formulation for a fairly simple tire compound.

Often side reactions or unexpected consequences occur when adding new materials to such a complex formulation. For example, soda lime borosilicate HGMs can have a moderately high pH of 9.5—10. Yet the formulation in Table 4.2 already has a fairly basic oxide additive—zinc oxide, used in part to accelerate and control the rate of the cross-linking reaction. It acts as an acid absorber balancer to the acidic sulfur vulcanizing agents. Adding another fairly basic additive to this mixture may change this delicate balance and cause scorch (too fast a viscosity build) or poor mill-ability (too slow of viscosity build). Commonly, when adding HGMs to rubber compositions, it is recommended to test cure rates through the use of an MDR rheometer to evaluate the potential effect of the pH on viscosity.

Table 4.2 Typical Tire Tread Compound Showing the Complexity of Additives Used in These Types of Formulations

Component	PHR
Natural rubber	75
Polybutadiene rubber	25
Carbon black	30
Silica	25
Silane coupling agent	2
Oil	25
Wax	1
Antioxidant #1	1
Antioxidant #2	1
Stearic acid	2
Zinc oxide	3
Vulcanization accelerator #1	1.5
Vulcanization accelerator #2	1

Application of HGM in Rubber; Example 1—Pneumatic Tires

A tire is made of a composite of a polymer or polymer blend with additives and fillers along with reinforcements that enable it to be cured into a finished part with a range of properties. As shown in Figure 4.9, there are three main parts of a tire; the tread, the sidewall, and the bead (the part that forms the connection to the rim). It is normal for each of these parts to have significantly different compounds even to the point of having different base rubber materials altogether. The objective is to tailor the needs of a compound to the requirements where they will be used in the tire. The whole tire is intended to be a complete composite article that has an incredible balance and range of properties and responses. For example, tire tread compounds need to be soft and pliable enough to provide good traction, but if they are too soft, they provide poor rolling resistance and generally show higher wear. On the other hand, a compound that is stiff may have better rolling resistance, but poorer

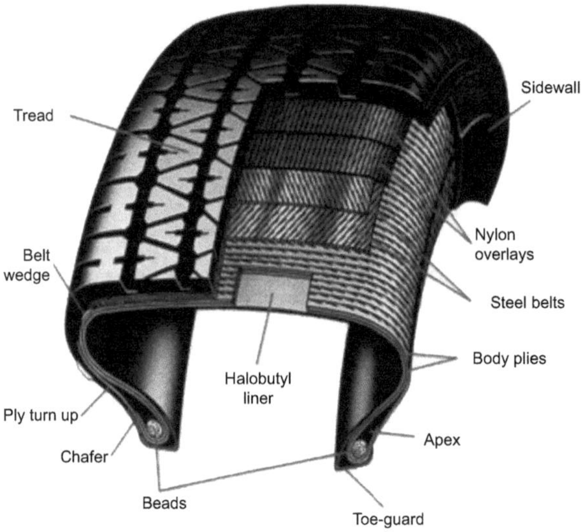

Figure 4.9 Tire design cross-section.

overall traction, especially wet traction. Tire engineers try to use both material properties and design to optimize for a specific set of application demands (winter vs all year for example).

Considering the forces at work on a tire it is almost inconceivable that an article can be made to function and function so smoothly for such a long time in this use environment. A tire is inflated so it has extensional and tensile forces along the sidewall. It has the weight of a car (or ¼ of the weight) bearing down on it causing compression. As it moves down the road surface, a surface that can be very hot and abrasive, it heats up due to frictional forces, expanding the gas in the interior causing more pressure to build up. Friction causes wear and dimensional change on the tread often concentrating stresses. It has to endure high impacts and lacerations (cut resistance) caused by pot holes and road debris. It is exposed to many, potentially extreme, elements—sun, wind, rain, snow, and ice. And hard breaking can expose the surface to grinding and tearing. Adding fillers like HGMs to such a dynamic system requires careful evaluation of physical property changes.

There are several patents that provide a glimpse into the weight reduction potential, physical property changes, and formulation optimization in pneumatic tires.

US patent number 7406990, assigned to Goodyear, characterizes the use of 0.60 g/cc specific gravity HGMs with 18KPSI isostatic crush strength in a run flat tire sidewall insert for weight reduction [1].

One example shows 5 PHR of HGM added to a 50/50 blend of natural rubber and polybutadiene. This formulation provides a little more than 2% weight reduction with no measurable microsphere breakage (process of incorporation is not specified) and essentially no effect on physical properties. A second example shows 5 PHR of HGM added to an 80/20 blend of natural rubber and polybutadiene respectively. Again this formulation provides slightly more than 2% weight reduction with no microsphere breakage. It does report a slight decrease in G′and a slight increase for hardness for the HGM containing formulation.

US patent application US2009255613, also assigned to Goodyear, provides a tread compound where similar high strength HGMs are used [2]. The tread is composed of three distinct compounds: a cap, a mid layer, and an inner layer. Concentrating the microspheres in the mid layer and coupling them to the rubber with a silane coupling agent provided the best physical properties and wear performance while affording 2−5 % weight reduction.

Application of HGM in Rubber; Example 2—Shoe Soles

A more recent application of HGM in rubber is the use in shoe soles to provide light weighting. A shoe consists of six main parts—the out, in, and midsole as well as the upper, heel, and shank. The common weight fraction of these parts for a walking shoe is shown in Table 4.3 sorted from high to low percent. When considering a weight reduction strategy, it is always good to know the weight fraction of the components of an article and target light weight technologies toward the highest weight fraction. For the components that are usually made from plastics and rubber the outsole has the highest weight fraction.

A recent Korean patent assigned to Kolon Industries Co., Ltd (KR20120021903) provides a good comparison of outsole compositions with and without HGMs [3]. In the examples they tested a 110 MPa crush strength HGM in several blends of different rubber materials to find an optimal formulation for HGM survival (lowest density of the cured compound) and best physical properties such as hardness, tensile strength, tear strength, and NBS abrasion. Some of the data in the patent is reprised in Table 4.4 showing a control outsole compound with a density of 1.11 g/cc and accepted physical properties for the application. A series of three HGM formulated examples show how reformulation (with respect to the 100 parts of base rubber) and concentration of the HGM can affect the

Table 4.3 Shoe Components and Weight Fraction of the Components

Component	Weight (g)	Weight Fraction (%)
Upper	128	31.7
Outsole	110	27.2
Midsole	60	14.9
Insole	47	11.6
Heel	28	6.9
Shank	31	7.7
Total	404	100

Table 4.4 Comparison of Shoe Sole Compositions With and Without HGM for Weight Reduction and Physical Properties

Component	Control	Formula 1	Formula 2	Formula 3
Butadiene rubber	30	70	70	64
Natural rubber	50	30	–	20
Syndiotactic polybutadiene	–	–	30	16
SBR	20	–	–	–
HGM	–	10	7	5
Specific gravity (g/cc)	1.11	0.935	0.925	0.910
Hardness (shore A)	70	67	71	63
Tensile strength (kgf/cm^2)	145	61.4	64.5	121
Tear strength (kgf/cm)	52	32.8	40.7	40
NBS abrasion (%)	250	163.8	181	287

HGM, hollow glass microspheres.

physical properties. Formulas 1 and 2 show good density/weight reduction but significant drop in tensile strength and tear strength. These compounds have exchanged SBR and natural rubber in the control for butadiene rubber. Formula 4 has excellent density reduction and physical properties, especially tensile strength and abrasion resistance, with good retention of tear and hardness. In this example a blend of three rubbers—butadiene, natural rubber, and syndiotactic butadiene rubber along with a lower HGM concentration seem to provide the optimal formulation.

A recent 3M Company patent application (WO201404455A1) details a way to impart HGMs from a masterbatch [4]. The masterbatch uses a blend of 1,4-*cis*-polybutadiene rubber and syndiotactic 1,2 polybutadiene and approximately 40 wt% of the HGM. This blend was arrived at after evaluating many rubber systems—NBR, BR, NR, SBR—and blends of rubbers with different mooney viscosity values maximizing HGM survival in the masterbatch. It also disperses readily in the base compound and allows for use in any mixer, extruder, and molding process with minimal breakage. In the examples it is let down to deliver around 6 wt% HGM (0.42 g/cc, 52 MPa) in the final shoe formulation to provide about 6% weight reduction to the outsole with good physical property retention such as abrasion and tear strength.

Application of HGM in Rubber; Example 3—Wire and Cable Compounds

Wire and cable constructions utilize plastics and rubber for jacketing and filling compounds due to their excellent electrical resistance/insulation properties as well as the ability to impart other attributes. Sometimes there exists a need for light weighting of these materials such as in transportation applications. Many new aircraft are "fly-by-wire" meaning they have replaced mechanical tension cables with the use of electrical signals, powering servo motors, to move control surfaces. These commercial passenger aircraft can have many miles of insulated electrical wire and cable. Reducing the weight of these compounds can provide significant weight reduction for the aircraft as a whole and reduce fuel consumption and CO_2 emissions. Use of rubber in these and other cable constructions is often for toughening cable jackets to provide tear resistance while dragging long cables through pipes and conduit (or over the ground).

Buoyancy is another example of how HGMs are used in elastomeric cable compounds. The use of HGMs in a TPE to provide "buoyancy-adjusting" cable jacketing and insulation between coaxial cables was described [5]. Buoyancy is often required in oil and gas drilling applications to counter the weight of subsea equipment attached to the floating platform. Without counter balancing the buoyancy of the equipment the platform would not be stable and would submerge. This patent describes the use of extremely high volume loadings (up to 70%) of 0.35–0.6 g/cc HGMs to provide 0.85 g/cc TPE jackets providing significant buoyancy in +1.0 g/cc seawater environments. Improved dielectric/insulation properties are obtained for the intermediate layer between the coaxial conductors as well.

The above three application examples are an attempt to provide a broad view of the application of HGMs in rubber. As described in these and other examples, HGMs can be used to reduce weight and improve insulation and physical properties. Many other applications in rubber exist. And still others are yet to be discovered.

References

[1] A.D. Brown et al., US 7,406,990 B2, August 5, 2008.
[2] P. Zhang, US20090255613 A1, 15 out, 2009.
[3] S.B. Han, C.S. Yoo. KR20120021903. Composition for Outsole and Shoes Manufactured by the Same.
[4] WO201404455A1.
[5] S. Kirkham, US patent 6096804, September 4, 1997.

5 Hollow Glass Microspheres in Sheet Molding Compounds

Baris Yalcin

Sheet Molding Compound Basics

Sheet molding compound (SMC) is a high-strength composite material comprising primarily a thermosetting resin, filler(s), and fiber reinforcement [1]. Thermosetting resin is typically based on unsaturated polyester (UP), vinyl ester (VE), phenolic or a modified vinyl urethane. Typical fillers are calcium carbonate (reduced cost), clay (improved surface), alumina trihydrate (fire retardance), talc (improved temperature resistance), mica (improved weathering), and hollow glass microspheres (weight reduction, thermal insulation). Most typical fiber reinforcement is glass but carbon and aramid fibers are also used especially in the more rigorous end uses, such as in military or aerospace applications. Other SMC ingredients used to enhance a performance or aid in the processing include low-profile additives (LPA), cure initiators, thickeners, process additives, and mold release agents. These are summarized in Table 5.1.

UPs based on maleic or fumaric acid were known since the 1920s. It was not until the invention by Carleton Ellis in 1936–1940 which combined UPs with styrene that led to the modern type of UP resins (UPRs) with a low viscosity, a high reactivity, and good material properties after cure. VE resins and shrink control additives appeared later in the 1960s and 1970s. Later vinyl urethane resins were introduced to the market.

UPR is prepared through a classical esterification process, which involves melt condensation of dicarbonic acids and glycols in a batch process at 190–220 °C. Water produced during this condensation reaction (few hours to 30 h) is removed by the elevated temperatures of the reaction along with vacuum and nitrogen stripping.

Raw materials for UPR are as follows:

1. Unsaturated dicarbonic acids/anhydrides to achieve crosslinking with styrene (e.g., maleic anhydride and/or fumaric acid).

Hollow Glass Microspheres for Plastics, Elastomers, and Adhesives Compounds
http://dx.doi.org/10.1016/B978-1-4557-7443-2.00005-0

Table 5.1 SMC Ingredients and Material Examples

Phase	Ingredients	Example
Paste a side	Unsaturated polyester resin	Orthophthalic unsaturated polyester resin with a styrene monomer content of 30–37 wt%
	Low-profile additive	Polyvinyl acetate with a styrene monomer content of 60 wt%, styrene–butadiene block copolymer solution with a styrene monomer content of 70 wt%
	Diluent	Styrene monomer
	Wetting/dispersing additive (process additive)	Phosphoric acid polyester copolymer
	Inhibitor	para-benzoquinone
	Internal lubricant (mold release)	Zinc stearate
	Cure initiators	*t*-Butyl peroxyisopropyl carbonate, *t*-butyl peroxybenzoate
	Filler(s)	$CaCO_3$, hollow glass microspheres, talc, ATH, mica etc.
Paste B side	Thickeners (maturating agent)	MgO-based thickening agent
Reinforcement	Fiber(s)	Glass fiber, carbon fiber

SMC, sheet molding compound.

2. Saturated dicarbonic acids/anhydrides to control crosslink density and to optimize network properties (e.g., phthalic acid anhydride, isophthalic acid, terephthalic acid).
3. glycols such as 1,2 propylenglycol (PG), dipropylenglycol (DPG), diethylenglycol (DEG).
4. Acids and glycols contributing special effects (dicyclopentadiene (DCPD)-capped resins).

It is difficult to classify UPRs by their chemical structures due to a large variety of raw materials and their combinations. Some of the raw material attributes are listed in Table 5.2. As a general guideline, there are three basic types of UPRs. However, within each of these families, there are hundreds of variants that incorporate these various chemical constituents

Table 5.2 Chemistry of Raw Materials and Their Attributes

Anhydrides/Acids	**Attribute**
Phthalic anhydride	Low cost, styrene compatibility
Maleic anhydride	Chemical resistance, rigidity
Adipic acid	Flexibility, toughness
Isophthalic acid	Toughness, chemical resistance, weathering resistance
Terephthalic acid	Higher heat resistance
Fumaric acid	Helps isomerization of maleic anhydride and limits acidity of reaction
Glycols	
Propylene glycol	Styrene compatibility
Ethylene glycol	Low cost, rigidity
Dipropylene glycol	Flexibility, toughness
Diethylene glycol	Flexibility, toughness
Methylpropane diol	Toughness, chemical resistance
Neopentyl glycol	UV and chemical resistance
Other	
Dicyclopentadiene	Low cost, shrinkage control

in different combinations and permutations to achieve the desired results depending on performance requirements and cost.

1. Orthophthalic resins

These are referred to as "general purpose UPRs" formed by maleic anhydride/phthalic anhydride/glycol. General-purpose UPRs exhibit sufficient mechanical properties but are prone to hydrolysis by water.

2. Isophthalic resins

These are formed by maleic anhydride/isophthalic acid/glycol. Replacing the phthalic anhydride with isophthalic acid reduces the sterically crowded esters and a high-strain energy allowing the cured product to be more resistant to hydrolysis. With the iso-UPRs, applications that require water resistance were made possible.

3. The DCPD resins

In this category, a large amount of bulky and highly aliphatic DCPD moiety is incorporated into a UPR polymer achieving a very cost-effective UPR and reduced shrinkage during curing. These two factors make such resins ideal for uses such as boat construction and tub and shower applications. Although DCPD resins are very brittle due to their ultra-low molecular weights, they are apparently adequate for these "not very demanding" applications.

Now that we briefly reviewed how UPRs are made, the next step is combining and mixing with styrene. Once the UPs are prepared and the condensation reaction has a turnover of 90–97%, the hot polyester melt is continuously fed to the relatively colder styrene and mixed in a second reactor whereby the temperature of the mixture is kept under 80 °C by cooling. The optimum mixing ratio of reactive double bonds to styrene is from 1:2 to 1:2.5 (on a molar basis) for most properties. With the addition of styrene, and in the presence of a catalyst and accelerator, the UPRs are cross-linked (cured) to form a three-dimensional network depicted in Figure 5.1. Note that styrene has two functions, that is, they reduce viscosity so the resins can be processed, and they cross-link with the double bonds in the polyester.

There are two main SMC categories, structural (non Class-A) and Class-A, which are influenced by formulation and for the most part by the amount of fibers. Glass fibers in SMC are chopped into lengths of

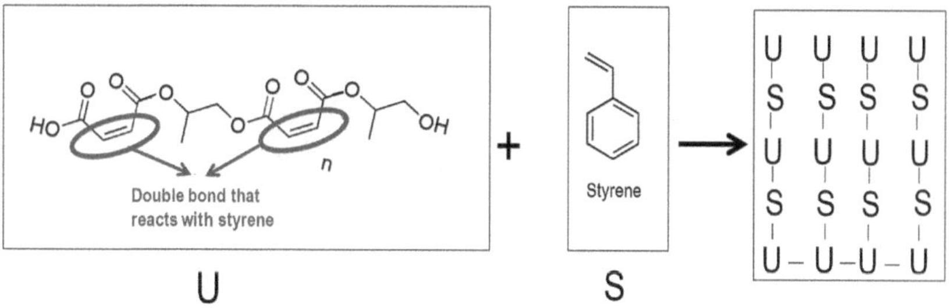

Figure 5.1 Styrene and unsaturated polyester curing schematics.

12–50 mm and the amount can vary from 25% to 60% by weight depending on the performance requirements, surface finish, processibility, and cost.

Structural SMC(s) have superior mechanical properties with higher glass fiber loadings than Class-A SMCs but exhibit poor surface finish and are typically targeted for nonvisible load-bearing parts. Class-A SMC(s), on the other hand, require superior surface finish with high aesthetics but typically exhibit lower mechanical properties due to lower glass fiber loadings. For Class-A surfaces, the overall percentage of glass fibers is typically limited to less than 30% by weight to reduce fiber read through and to optimize surface smoothness. Other forms of glass fibers such as continuous strands, woven rovings, and glass cloths are also avoided for surface smoothness in Class-A applications.

For Class-A surface requirements, in addition to low glass fiber amount, LPA are used to control shrinkage as polyester resin polymerizes or cures. SMC shrinkage is usually measured as a percentage difference between the measurement of the mold and a cured part, both taken at room temperature. A typical non Class-A SMC compound exhibits an apparent shrinkage greater than 0.05% while a low-profile SMC will shrink less than 0.05%. LPA are generally materials such as poly(vinyl acetate), polystyrene, polyethylene, or polycarbonate. During the UP cure cycle, the LPA separate into a second phase, which expands to counteract the shrinkage of the curing UPR. The science of LPA significantly contributed to the expansion of the SMC markets into exterior automotive markets where Class-A quality surfaces is a requirement. When using Class-A formulations, part-design techniques in addition to material formulation play a crucial role to improve mechanical properties. Figure 5.2 shows representative weight and volume

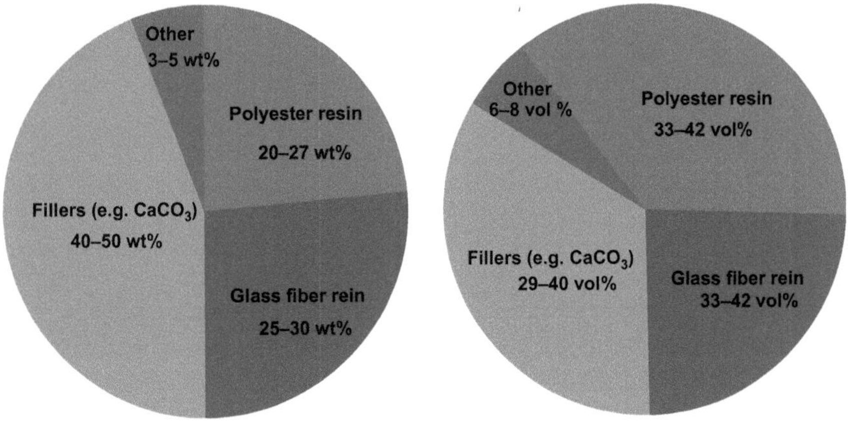

Figure 5.2 Sheet molding compound primary components representative weight and volume percentages for Class-A applications composition.

percentages of glass fibers with respect to other primary components in a low-profile Class-A formulation.

SMC Process

The process used for SMC manufacturing is illustrated in Figure 5.3. All ingredients except the glass fibers are mixed together to form a resin slurry (or paste), which has a viscosity of about 20,000–40,000 mPa s (consistency of molasses or honey). Resin paste A-side includes UPR, LPA, diluents (styrene monomer), wetting and dispersing additives, peroxide initiator (catalyst), mold release agent, and fillers (typically $CaCO_3$). Resin paste B-Side which typically contains an alkaline earth thickener material, monomer, and a carrier resin is combined which is mixed with the A-side paste in the compounding area at a ratio of 32–35:1. This mixture of A and B sides (which is referred to as A/B paste) is cast onto a thin plastic carrier film moving under a doctor blade, which controls the amount of the resin slurry applied and spreads it evenly. Concurrently, glass fiber rovings are fed into a cutter and chopped fibers are dropped onto the resin paste. Downstream from the chopping operation, a second carrier film is coated with A/B resin paste and is laid on top of the first carrier film creating a resin paste and glass fiber sandwich. The sandwich film structure is passed through a set of compaction rollers, which knead the fibers into the resin squeezing

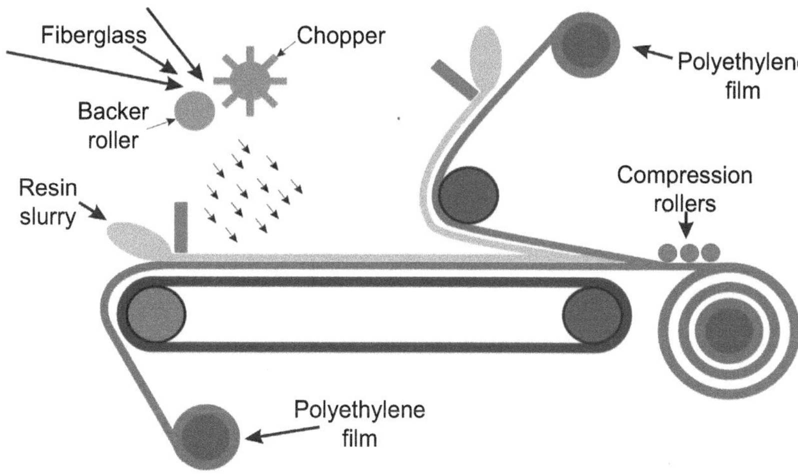

Figure 5.3 Sheet molding compound process.

excess trapped air out of the sheet for uniform wetting. The compound sandwiched between the carrier films is gathered into rolls and stored until it matures. The carrier film used for sheet casting is usually a multilayered polyethylene film with a polyamide core. Polyamide core is useful to prevent styrene monomer permeation during storage whereas the polyethylene layers help with the release of the SMC upon maturation.

Maturation time is required to allow the relatively low-viscosity resin to thicken chemically. Thickeners (B-Side resin) provide ionic bonding of alkaline earth oxides with acid functionality in the resin and help control maturation. The thickened SMC is easier to handle and prevents the resin paste from being squeezed out of the fibers. Upon maturation (3–5 days), SMC is tack free and reaches a viscosity of 40–100 million mPa s (leather-like consistency). The carrier film is removed and the sheet is prepared into a charge of predetermined weight and shape. The charge is placed on the bottom of two mold halves in a compression press.

Hollow Glass Microspheres in SMCs

There are several inorganic fillers used in SMC formulations and hollow glass microspheres fall into the category of inorganic filler additives in SMCs. Benefits of hollow glass microspheres in SMC include the following:

1. Weight reduction of up to 40% lower than the 1.8−1.9 g/cc density of industry-standard SMC formulations.
2. Improved dimensional stability.
3. Improved thermal insulation.
4. Reduced dielectric constant.
5. Transparent to microwave radiation.
6. Improved stiffness at a given weight.

Industry-accepted SMC formulations, with a density of 1.8 g/cc, contain approximately 40 wt% (27 vol%) calcium carbonate. The primary function of using high calcium carbonate loadings is to reduce the amount of resin in the system to control price and specific process and performance attributes such as consistency and providing a smoother surface on the final product. In addition, they absorb some of the heat of the curing reaction, and also lessen internal strains and settling effects due to the extreme viscosity changes, which might cause a more porous surface. The bulk fillers such as $CaCO_3$ also tend to reduce thermal expansion and shrinkage. Calcium carbonate has functioned for many years as important filler in automotive SMC formulations. Due to the recent mandates in CAFÉ standards, however, SMC manufacturers have adopted various venues to reduce weight of their SMC formulations one of which is replacing heavy fillers with hollow glass microspheres. Calcium carbonate has a density of 2.7 g/cc and increases the weight of a finished part since UPRs have an approximate density of 1.0 g/cc. The use of hollow glass microspheres densities has allowed SMC formulations to achieve densities as low as 1.2 g/cc, ~40% lower than industry standard.

When formulating with hollow glass microspheres in SMCs, the following parameters should be considered to maintain an acceptable balance of weight, surface aesthetics, mechanical performance, and processing characteristics.

1. Correct amount (% loading)
2. The right choice of GB collapse strength
3. Particle size of the hollow glass microspheres.

The right choice of hollow glass microsphere collapse strength is critical during processing, mainly during paste mixing and molding.

Recommended mixers are horizontal plow mixers, ribbon mixers (e.g., Littleford, Baker perkins and etc.), or any low/medium shear mixer. Cowles mixers can also be used but with caution, at low speed and limited mixing time. Hollow glass microspheres should be added last after all other ingredients are mixed thoroughly. Since hollow glass microspheres are low-density fillers, it is also important to maintain mild agitation in mix tanks to prevent float-out. Thixotropes are also used to prevent hollow glass microspheres floating to the surface of the paste mixture.

When transferring SMC formulation containing hollow glass microspheres, gear pumps should be avoided and instead, low shear progressive cavity (Moyno), double diaphragm, or lobe pumps should be used. It is also recommended that the working pressure of the pump does not exceed the isostatic pressure rating of the hollow glass microsphere being used.

During molding, maximum clamping force should not exceed the isostatic pressure rating of selected hollow glass microsphere. Mold closing speed should be as slow as practical during the final stage of the mold close to allow the SMC charge time to fill the mold cavity without building undue pressure against the bubble filled charge. Undue pressure can lead to high-density parts and short shots.

Hollow glass microspheres can replace all or part of the bulk fillers, for example, $CaCO_3$, used in SMC formulations depending on the weight reduction requirements. However, for a starting formula, the general rule of thumb is to replace half of the $CaCO_3$ volume with same volume of hollow glass microspheres and additionally 25−35%. For example, 45 vol % $CaCO_3$ in the original formula would be replaced by 22.5 vol% $(45 \div 2)$ $CaCO_3$ and 28 vol% $((45 \div 2) + 0.25 \times (45 \div 2))$ hollow glass microspheres in the reformulated version. Note that the total filler volume content would, in this case, be increased from 45 vol% to 50 vol %. Due to the spherical nature and low resin demand of hollow microspheres, such an increase in total volume content of inorganic filler does not adversely affect viscosity while reducing resin content. Surface area (BET), particle size distribution, and oil absorption are all $CaCO_3$ parameters that influence the degree of hollow glass microspheres usage in the formula. SMC formulations containing hollow glass microspheres may require more thickener, that is, maturation agent (magnesium oxide dispersion) than a heavyweight formulation of similar composition to achieve same thickening rate. The actual amount of increase in thickener depends on (1) the resin system employed, (2) the thickening rate desired (e.g., 3-day maturation vs. 2 weeks), (3) intended mold coverage, and

Table 5.3 SMC Paste Formulation With $CaCO_3$

Heavy Weight Formula					
Raw material	**Density (g/cc)**	**Weight (kg)**	**Wt%**	**Volume (l)**	**Vol%**
UP resin	1.13	60.00	18.43	53.10	29.78
Low-profile additive	1.01	40.00	12.29	39.60	22.21
Catalyst	1.04	1.50	0.46	1.44	0.81
			0.00	0.00	0.00
MgO paste	1.70	4.00	1.23	2.35	1.32
			0.00	0.00	0.00
Zinc stearate	2.30	5.00	1.54	2.17	1.22
$CaCO_3$	2.70	215.00	66.05	79.63	44.66
Totals	1.826	325.50	100.00	178.30	100.00

(4) the desired molding viscosity of the SMC. The rule of thumb for a starting point is to increase thickener 2% by volume for every 1 vol% hollow glass microsphere added to the formula. For instance, if we assume a standard SMC formulation with ~45 vol% $CaCO_3$ as shown in Table 5.3 and replace half of the total volume of $CaCO_3$ with about 135% hollow glass microsphere of density 0.38 g/cc as shown in Table 5.4, we would be using ~22.3 vol% $CaCO_3$ and 30 vol% hollow glass microspheres (density 0.38 g/cc). This would require increasing the thickener amount by 60% by volume (2 × 30 vol% GB) in the low-density SMC formula where the total batch volume is not adjusted as shown in Table 5.5. Comparing Tables 5.3 and 5.5, we can see that the absolute volume in liters has been increased from 2.35 l to 3.76 l (3.76/2.35 = 1.6).

Hollow glass microspheres do not serve as reinforcement in the formulation, which is why, when adding spheres, the amount of glass fiber by volume should stay the same to maintain similar net performance for the lighter weight paste.

Table 5.6 shows several SMC formulations including low density—low profile (LD-LP) and Low-density flame retardant SMC. To get an SMC with a Class-A surface, two different mixtures were used for each doctor

Table 5.4 Low-density SMC Paste Formulation With $CaCO_3$ and Hollow Glass Microspheres Where the Total Absolute Volume is Kept Constant as in the Original Formula (178.30 l) of Table 5.3

Low-Density Formula, Same Total Batch Volume					
Raw Material	**Density (g/cc)**	**Weight**	**Wt%**	**Volume**	**Vol%**
UP resin	1.13	51.00	22.82	45.13	25.31
Low-profile additive	1.01	34.00	15.22	33.66	18.88
Catalyst	1.04	1.28	0.57	1.23	0.69
			0.00	0.00	0.00
MgO paste	1.70	5.44	2.43	3.20	1.79
			0.00	0.00	0.00
Zinc stearate	2.30	5.00	2.24	2.17	1.22
$CaCO_3$	2.70	106.42	47.62	39.41	22.11
			0.00	0.00	0.00
HGM (3 M glass bubble S38)	0.38	20.33	9.10	53.49	30.00
Totals	1.253	223.46	100.00	178.30	100.00

blade, one formulation with no or very low content of hollow glass microspheres (HGMs), the other with a high amount of microspheres. Before molding a part, the two layers are put together in such a way that the HGM-filled sides of the SMC lay on the inside of the charge as shown in Figure 5.4 [2].

The use of sandwich construction is a successful technique to reduce the density of the SMC parts using larger particle size hollow glass microspheres, typically >40 μm, while maintaining Class-A surface quality. Unless used in sandwich construction form, hollow glass microspheres with larger diameters do not yield Class-A appearance. This is because sanding, which is a critical operation before painting used to increase the surface bonding between the primer and the paint, has been associated with generating paint pops exasperated by the large particle

Table 5.5 Low-Density SMC Paste Formulation With $CaCO_3$ and Hollow Glass Microspheres Where the Total Absolute Volume is Not Adjusted and the Weight of UP Resin, Low-Profile Additive, and Catalyst is Kept Constant

Low-Density Paste Formula, Total Batch Volume Not Adjusted					
Raw Material	**Density (g/cc)**	**Weight**	**Wt%**	**Volume**	**Vol%**
UP resin	1.13	60.00	22.82	53.10	25.31
Low-profile additive	1.01	40.00	15.22	39.60	18.88
Catalyst	1.04	1.50	0.57	1.44	0.69
			0.00	0.00	0.00
MgO paste	1.70	6.40	2.43	3.76	1.79
			0.00	0.00	0.00
Zinc stearate	2.30	5.88	2.24	2.56	1.22
$CaCO_3$	2.70	125.20	47.62	46.37	22.11
			0.00	0.00	0.00
HGM (3 M glass bubble S38)	0.38	23.91	9.10	62.93	30.00
Totals	1.253	262.89	100.00	209.76	100.00

size fillers. It is theorized that sanding breaks the portion of hollow glass microsphere that are close to the molded part surface. This creates depressions in the surface, which traps the low vapor pressure solvents used in paint primers. During the subsequent paint operations, the temperature required to drive off higher vapor pressure solvent in paint top coats would cause the retained primer solvent to vaporize also.

When sandwich construction cannot be used, hollow glass microspheres with smaller particle size and good bonding to the polyester enables Class-A appearance [3].

US 5,412,003 [4] pointed out the importance of elution alkalinity of HGMs in producing lightweight SMC and BMC molding compounds with superior workability and paintability. It was reported that an elution alkalinity of 0.05 meq/g or less was critical to producing compounds with

Table 5.6 Example Low-Density SMC Formulations [2]

Component	Standard Low-Profile Class-A SMC	Low Density—Low Profile (LD—LP) Class-A SMC Doctor Box-I	Low Density—Low Profile (LD—LP) Class-A SMC Doctor Box-2	Low-Density Flame Retardant SMC	Ultra-Low Density SMC
UP resin (kg)	60	60	60	80	80
Polyvinylacrylate solution (kg)	40	40	40	20	—
Polystyrene solution (kg)	—	—	—	—	20
Styrene (kg)	—	—	10	10	10
BYK®- W 972 (kg)	0.5–1	—	0.5–1	0.75	0.5–1
BYK®- W 972 (kg)	2	—	5	6	6
Tertiary butylperbenzoate (TBPB) (kg)	1.50	1.50	1.50	1.50	1.50
Internal release agent (kg)	4	5	5	6	5
Hollow glass microsphere (kg) (0.37 g/cc)	—	—	55	20	65

(*Continued*)

Table 5.6 Example Low-Density SMC Formulations [2] (*Continued*)

Component	Standard Low-Profile Class-A SMC	Low Density –Low Profile (LD–LP) Class-A SMC Doctor Box-I	Low Density –Low Profile (LD–LP) Class-A SMC Doctor Box-2	Low-Density Flame Retardant SMC	Ultra-Low Density SMC
Calcium carbonate	200	100	20	–	30
Aluminum trihydrate (ATH)-21 μm (kg)	–	–	–	200	–
Aluminum trihydrate (ATH)-2 μm (kg)	–	–	–	80	–
Pigment Paste (kg)	–	–	–	7	–
Magnesium oxide Paste (kg)	3	3.5	6	–	4
Magnesium hydroxide Paste (kg)	–	–	–	6	–
Glass roving, 25 mm length (%w/w)	25	38	38	27	36
Glass roving, 25 mm length (%v/v)	19	19	19	17	15
Density (g/cm^3)	1.85	1.3	1.3	1.7	1.1

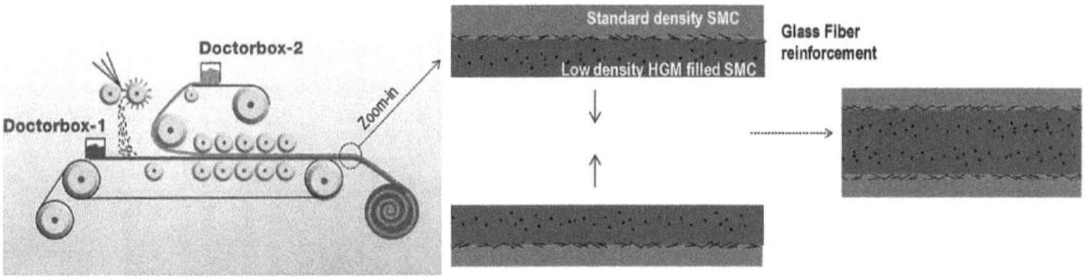

Figure 5.4 Sandwich low-density Class-A sheet molding compound construction setup. *Image reconstructed from SAE 1999-01-0980.*

superior paintability and adhesion with no blisters. Table 5.7 shows the compositions and results where no stickiness and good workability are indicated by "A," those which are workable although with some stickiness are indicated by "B," and those which are strongly sticky and difficult to work with are indicated by "C." Each SMC was compression-molded at a pressure of 80 kgf/cm^2 and a flow time of 9 s to produce flat-panels (300 × 300 × 2 mm). In melamine coating, melamine alkyd paint was coated, and baked at 140 °C. In urethane coating, urethane paint was coated and baked at 80 °C. After each coating, SMC panel was soaked in warm water of 40 °C for 10 days, the number of blisters formed on the coated surface was counted, and adhesiveness was evaluated by a crosscut adhesion test.

Recently, DSM, a major supplier of synthetic resins used in SMC systems, reported reduced density SMCs ~1.3 g/cm^3 (30% reduction compared to conventional Class-A) with well-maintained surface quality and mechanical properties [5]. In order to achieve good mechanical performance with low density, they utilized a high-performance LPA along with silane-coated hollow glass microspheres to improve bond between the polyester and the microspheres.

The starting formulation used in the low-density "Class-A" SMC paste is shown in Table 5.8. The optimum filler to HGM ratio was 60:28 (on a volume basis) in order to have the right balance of SMC paste viscosity and processability, mechanical performance, and molded part surface quality.

Types of silanes used to promote adhesion of the HGS to the matrix are shown in Table 5.9.

The importance of using a high-performance LPA is apparent when compared to a conventional LPA as shown in Table 5.10 Further improvement in mechanical properties was achieved when silane-treated hollow glass microspheres were used as shown in Figures 5.5 and 5.6.

Table 5.7 Light Weight Unsaturated Polyester Compounds—Effect of Elution Alkalinity of HGMs [4]

	1	2	3	4	5
Component	**PHR**	**PHR**	**PHR**	**PHR**	**PHR**
Unsaturated polyester resin A	43	43	43	43	43
Styrene	7	7	7	7	7
Styrene/butadiene block copolymer solution	50	50	50	50	50
Polyvinyl acetate solution	–	–	–	–	–
l,l-bis(t-butylperoxy) 3,3,5~trimethyl cyclohexane	1	1	1	1	1
Hollow glass microsphere					
HGM- 0.50 g/cc; elution alkalinity = 0.020 meq/g	22	–	–	–	–
HGM- 0.50 g/cc, elution alkalinity = 0.050 meq/g	–	22	–	–	–
HGM- 0.50, elution alkalinity: 0.080 meq/g	–	–	22	–	–
HGM-1.10; elution alkalinity = 0.020 meq/g	–	–	–	70	
HGM-1.10, elution alkalinity = 0.075 meq/g	–	–	–	–	70
p-Benzoquinone	0.02	0.02	0.02	0.02	0.02
Calcium carbonate	100	100	100	60	60
Magnesium oxide	1	1	1	1	1

(Continued)

Table 5.7 Light Weight Unsaturated Polyester Compounds—Effect of Elution Alkalinity of HGMs [4] (*Continued*)

	1	2	3	4	5
Component	**PHR**	**PHR**	**PHR**	**PHR**	**PHR**
Glass fiber	28%	28%	28%	28%	28%
Viscosity (40 °C, Brookfield 2.5 rpm × 10,000 cPS)					
After 24 h	896	512	12	928	125
After 48 h	1376	864	145	1408	248
After 72 h	1856	1216	282	2144	369
Workability of SMC	A	A	C	A	C
Density of products molded from SMC (g/cc)	1.50	1.51	1.51	1.51	1.51
Water resistance of coated panels (after soaking for 10 days in warm water at 40 °C)*					
Melamine coating					
Adhesiveness*	0/100	0/100	15/100	0/100	13/100
No of blisters*	0	0	89	0	75
Urethane coating					
Adhesiveness*	0/100	0/100	45/100	0/100	32/100
No of blisters*	0	0	58	0	47

Adhesiveness by crosscut adhesion test, expressed by (number of peels)/100.
No. of blisters: number of blisters within molded and coated panel of 300 × 300 mm.

Table 5.8 Starting Formulation of Low-Density Class "A" SMC [5]

Component	(PHR)
Unsaturated polyester resin Palapreg® P 0423-02	65.5
Low-profile additive (LPA 1): Palapreg® H 2681-01 or Palapreg® H 2700-01	28
Low-profile additive (LPA 2): (Palapreg® H 1080-01)	2
Filler (calcium carbonate)	60
HGM (surface modified/non-modified) 0.38 g/cc D50:40 μm	28
Glass fiber	38%

Courtesy of DSM.

Table 5.9 Silane Types Used in the Experiments [5]

Types of Silane	Source	Code
Methacryloxypropyl trimethoxysilane	Momentive Silquest® A-174NT	a
Glycidoxypropyl trimethoxysilane	Momentive Silquest® A-187NT	b
Mercaptopropyl trimethoxy	ABCR chemicals AB 111219	c
Aminopropyl trimethoxysilane	Acros 313251000 ACD code: MFCD00008206	d

Courtesy of DSM.

Table 5.10 Flexural Strength and Modulus of Low-Density SMC Formulation Utilizing High-Performance and Conventional Low-Profile Additive (LPA) Using Untreated Hollow Glass Microspheres [5]

	Flexural Strength (MPa)	Flexural Modulus (MPa)
LPA Palapreg® H 2681-01 (conventional)	103	7100
LPA Palapreg® H 2700-01 (high performance)	138	7700

Reconstructed table–courtesy of DSM.

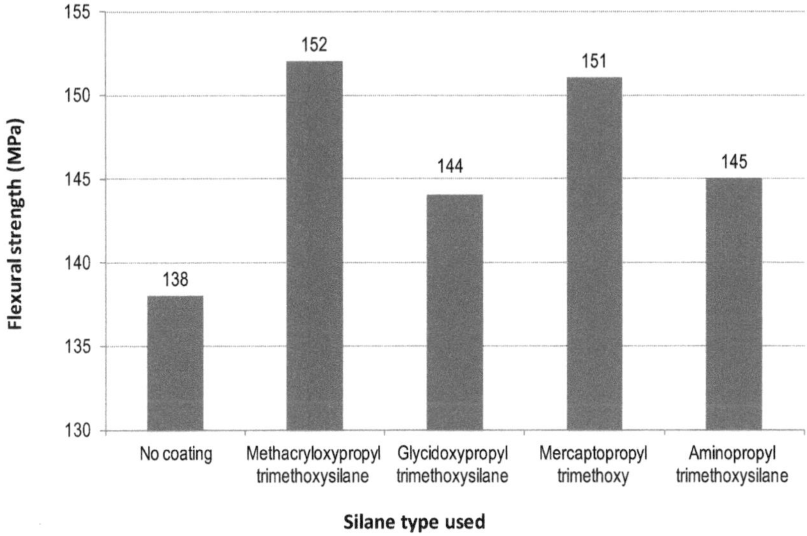

Figure 5.5 Flexural strength as a function of silane type used for HGS in low-density sheet molding compounds utilizing high-performance LPA (Palapreg® H 2700-01). LPA, low-profile additive. *Reconstructed from [5] courtesy of DSM.*

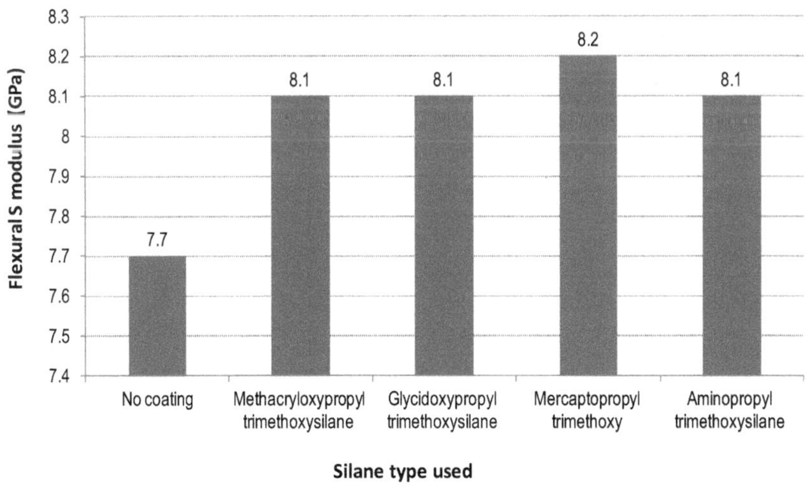

Figure 5.6 Flexural modulus as a function of silane type used for HGM in low-density sheet molding compounds utilizing high-performance low-profile additive (Palapreg® H 2700-01). *Reconstructed from [5] courtesy of DSM.*

Although this chapter has so far focused mainly on SMCs, one can consider bulk molding compounds (BMC), spray up layup, and resin transfer molding sister technologies, utilizing similar components and benefiting from the same attributes that HGMs impart to SMCs. In fact, the use of hollow glass microspheres in thermoset-laminating compositions along with fibrous mats of various forms was reported for polyesters, phenolics, epoxies, and silicons as early as in 1967 [6]. In the production of laminated thermoset composites, the hollow glass microspheres were mixed with the particular laminating thermoset resin to be employed by conventional mixing methods. If the laminating resin was in the liquid state, for instance, the hollow glass microspheres were combined with the laminating resin in a conventional shear-type mixer. If the laminating resin was in the powder state, as with phenolic powders, the hollow glass microspheres were mixed by simple salt and pepper intermixing. If the laminating resin was in solid state form, HGMs were added on a rubber mill, care being taken not to exceed the curing temperature of the resin. In order to form the laminates, alternating layers of HGM-containing resin and reinforcing fabric mats were placed in the mold and fused and solidified by heat and pressure. When the resin was initially a liquid, the reinforcing fabric mat was coated with the liquid resin and hollow glass spheres followed by another sheet of coated glass fabric superimposed (Table 5.11).

BMC, similar to SMC, consist of a thermosetting resin, glass fiber reinforcement, fillers as well as additional ingredients such as LPA, cure initiators, thickeners, mold release agent added to enhance performance

Table 5.11 HGM versus $CaCO_3$-Filled Laminated Plastic [6]

	HGM	$CaCO_3$
Wt% in laminate	2.1	25.2
Wt% reinforcing glass fiber	15	40
Density of laminate, g/cc	1.44	1.82
Tensile strength, psi	14,400	15,000
Flexural strength, psi	26,900	2500
Flexural modulus, psi	1.74×10^6	1.30×10^6
Wet strength retention, flexure, psi	19,700	17,500
Water absorption, wt%	0.6	0.5

of processing of the material. All liquid ingredients (including all the cure initiators, mold release agent and others) and some powder are mixed separately. In a mixer, this paste plus the remaining filler and the chopped glass fibers are homogenized and stored in plastic bags. The material needs at least 7 days maturation to attain all its properties. BMC is transformed from its liquid and fiber ingredient into a bulk product that can be squeezed up into cylinder-like shapes. BMC features more or less similar properties to SMC. Due to shorter fiber length, it flows easily into the smallest cavities. BMCs are typically processed via injection molding whereby the BMC is injected into a hot mold (160 °C). Curing speed is faster than with the compression molding process.

Besides UPs, other thermoset resins have also been successfully used with hollow glass microspheres. For instance, low-shrinkage phenolic molding compound suitable for SMC with improved maturation times was reported in US Patent Number: 4,794,051 [7]. The compound contained resole phenolic resin, curing agent, a blend of clay, talc, and hollow glass microsphere filler with a butyrolactone diluent and fiber reinforcement. Lactone-reactive diluent provided a reduced initial viscosity desired during the mixing stage, and an increased viscosity during the maturation stage. It also allowed a higher loading of fillers. In addition, the lactone in the phenolics molding composition resulted in higher impact strength and lower shrinkage. In the preparation of the compound resole phenolic resin, silane coupling agent and butyrolactone was mixed on a high-speed mixer. Mixing was continued while the filler mixture was added to the resin mixture over a period of about 3–5 min. Immediately following the mixing, the temperature of the resulting treating mix was measured and found to be about 60 °C. The treating mix was then further mixed with 30% glass fibers in a Baker Perkins mixer to form a bulk molding compound. The bulk molding compound was maturated at room temperature for several days and then compression molded as 8-inch square slabs at 325 °F and 500 psi with a curing cycle of 4 min. The shrinkage of the slabs was measured in reference to the cold dimensions of the mold. The slabs were cut and machined and tested for impact strength and flexural strength in accordance with ASTM standard methods with the results in Table 5.12.

A final word: HGMs will continue to be an important ingredient of UP formulations due to the lightweight benefits that they impart along with Class-A surface quality.

Table 5.12 Low-Shrinkage Phenolic Molding Compound Example Containing a Blend of Clay, HGMs, and Talc [7]

	1	**2**	**3**	**4**
Resol phenolic resin	510	510	510	600
Butyrolactone (reactive diluent)	90	90	90	–
Clay (filler)	125	162	200	125
Hollow glass microspheres (filler)	125	125	125	125
Talc (filler)	125	125	125	125
Coupling agent γ-aminopropyl triethoxy silane	36	36	36	36
$Ca(OH)_2$ (curing agent)	23.7	23.7	23.7	23.7
$Mg(OH)_2$ (curing agent)	23.7	23.7	23.7	23.7
Zinc stearate (mold release agent)	17.5	17.5	17.5	17.5
Fiberglass (1/2″)	375	375	375	375
Viscosity at 60 °C (cP)	12,800	13,400	25,600	59,200
Viscosity at RT (cP)	448,000	528,000	688,000	1,216,000
Viscosity after 3 days at RT (cP)	7,520,000	16,000,000	–	3,200,000
Shrinkage (inch/inch)	0.009	0.0011	0.0009	0.0021
Notched izod impact strength (ft-lb/inch)	1.94	1.71	1.92	1.43
Flexural strength (psi)	12,660	11,740	12,120	14,660

References

[1] Laurent Orgéas, P.J.J. Dumont, Sheet molding compounds, in: L. Nicolais, A. Borzacchiello (Eds.), Wiley Encyclopedia of Composites, second ed., John Wiley & Sons, Inc, 2012.
[2] B.V. Gregl, L.D. Larson, M. Sommer, J.R. Lemkie, Formulation Advancements in Hollow-Glass Microspheres Filled SMC. Society of Automotive Engineers. SAE-1999-01-0980.
[3] 3M Case Study. Supplier Reduces Weight of Sheet Molded Composite Plastic Parts by more than 25%.
[4] Akiyama, et al., US Patent 5412003 (May 1995).
[5] A. Hamarneh, B. Gorzolnik, A. Horbach, DSM proposes new roads to weight reduction, Press Information Schaffhausen (CH), November 21 2013.
[6] H.E. Alford, US patent 3316139 25 (April 1967).
[7] M.K. Gupta, US 4794051, Low shrinkage phenolic molding compositions (Dec 1988).

6 Hollow Glass Microspheres in Thermosets—Epoxy Syntactic Foams

Dinesh Pinisetty, Vasanth C. Shunmugasamy and Nikhil Gupta

Rapid growth in applications of syntactic foams in transportation and oil industry has motivated research and development activities in polymer matrix syntactic foams. While a wide range of matrix and particle materials have been used in fabricating syntactic foams, this chapter specifically focuses on the properties of epoxy matrix glass hollow particle-filled syntactic foams. Among mechanical properties, tensile, compressive, and flexural properties are discussed. In addition, dielectric and thermal properties are also discussed. Data on these properties have been extracted from all available studies and compared with respect to the syntactic foam density to identify the common trends. All these properties of syntactic foams can be tailored either by the hollow particle volume fraction or by the wall thickness. The possibility of using these two parameters independently enables developing syntactic foams with two or more properties tailored at the same time. Such possibility allows developing multifunctional syntactic foams and tailoring their properties for a wide range of applications.

Introduction

Low-density composite materials produced by random dispersion of hollow particles in a matrix material are defined as syntactic foams [1,2]. These foams are classified as closed-cell foams and particulate composites because the porosity in these materials is enclosed within hollow particles. The closed-pore structure gives advantages of low density, low moisture uptake, and excellent mechanical properties [3]. A combination of low density and moisture absorption makes syntactic foams suitable for fabricating floatation devices [4] and submarine buoyancy systems [5]. Syntactic foams are often used as core materials in sandwich composites because they ensure high rigidity and compressive strength of sandwich structures [6].

Hollow Glass Microspheres for Plastics, Elastomers, and Adhesives Compounds
http://dx.doi.org/10.1016/B978-1-4557-7443-2.00006-2

Syntactic foams are being explored as multifunctional composite materials due to their attractive mechanical, electrical, and thermal properties. A recent example of multifunctionality is the deckhouse of USS Zumwalt, where syntactic foam is used for stealth and structural strength. High dimensional stability and load-bearing capacity combined with thermal stability of syntactic foams make them attractive in aerospace, automotive, civil as well as marine structural applications compared to the conventional materials used in these applications.

Hollow particles play an important role in determining the properties of syntactic foams. The hollow particles of glass [7,8], carbon [9,10], fly ash cenospheres [11], ceramics such as Al_2O_3 [12] and SiC [13], and polymers [14–16], expandable perlite spheres [17] have been used in syntactic foams. Hollow glass particles are the most widely used fillers and are referred to by several terms in the literature, including glass microballoon and hollow glass microsphere (HGM). Glass microspheres provide a variety of properties to syntactic foams including low density, enhanced impact strength, improved dimensional stability, greater thermal insulation, and low dielectric constant. High-quality HGMs are now commercially available from several sources.

Some of the commonly used polymer matrix materials include epoxy [18], vinyl ester [19–21], cyanate ester [22], phenol [23], and bismaleamid [24]. One of the most widely used polymer matrix is the thermoset resin epoxy [18, 25–30]. Epoxy resin has high strength and stiffness, thermal and environmental stability, creep resistance, and lower shrinkage and water resistance. Epoxy/HGM syntactic foams are the most widely studied syntactic foams for mechanical, thermal, and electrical properties.

Application of Epoxy Syntactic Foams

Syntactic foams are finding applications in fields that are as diverse as deep-sea vehicles, space vehicles, aircraft, snow skis, soccer balls, and thermal insulation in underwater pipelines. A detailed list of current applications has been presented in a recent publication [31]. The main applications of syntactic foams can be divided into three categories based on the properties that they use:

1. Compressive properties: HGMs incorporate porosity in the structure of syntactic foams, while keeping the compressive strength high because the stiff glass particles become load-bearing elements

under compression. The compressive properties of syntactic foams are effectively used in submarines and underwater unmanned vehicles because of hydrostatic compression encountered by these structures. The National Oceanic and Atmospheric Administration (NOAA) uses Alvin and Jason underwater vehicles that are made of syntactic foams for deep-sea exploration. Among other examples is the Challenger craft used by James Cameron in a solo dive to the Mariana Trench in 2012. The main body of the craft was made of aramid fiber-reinforced syntactic foams.

2. Thermal insulation: Syntactic foams contain about 30–50 vol% voids formed by HGMs. The uniformly distributed void pockets and insulating nature of epoxy matrix provide insulating properties to syntactic foams. These materials are used as insulation in deep-sea oil and gas pipelines. External fuel tanks used in space vehicles are also insulated using syntactic foams.
3. Dimensional stability and low coefficient of thermal expansion (CTE): Syntactic foams are used in making tooling for composite materials. Low CTE of tooling leads to only small dimensional changes and distortions as the temperature changes during composite fabrication for reasons such as exothermic curing reactions. Low CTE makes it easy to maintain dimensional tolerances. In addition, syntactic foams are also easy to machine. Therefore, complex shapes can be machined for use as tooling. Low CTE and high dimensional stability makes syntactic foams suitable for tooling in thermoforming applications. An example of thermoforming die and tooling is shown in Figure 6.1. Thermoforming is a widely used process for large volume production of plastic parts such as disposable cups and carryout food containers.

Hollow Particle Properties

A representative sample of 3M HGMs is shown in Figure 6.2(a). These particles are spherical in shape and the particle size can vary over almost two orders of magnitude as observed in this figure. The average particle size, provided by the manufacturer, is presented in Table 6.1 for four types of HGMs widely used in manufacturing syntactic foams. Figure 6.2(b) shows that the walls of these particles are very thin relative to the particle diameter.

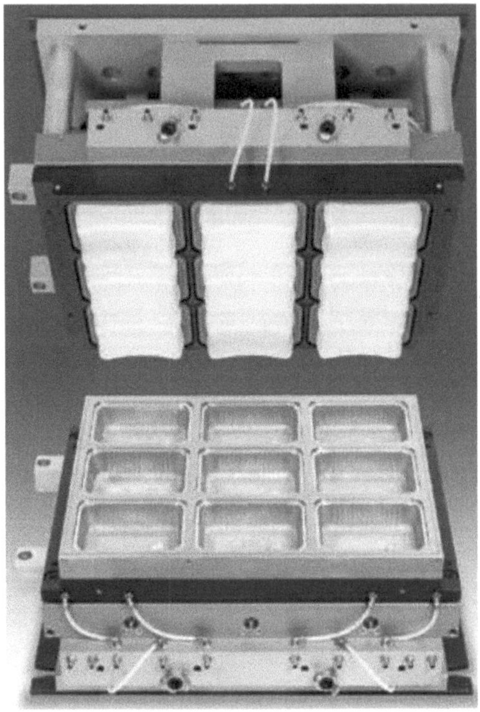

Figure 6.1 An example of plugs made of syntactic foams used in thermoforming. *Images courtesy Jeff Barker, Tooling Technology LLC, Fort Loramie, OH.*

The particle wall thickness can be related to a parameter named radius ratio η, defined as the ratio of inner radius to the external radius and given as

$$\eta = \frac{r_i}{r_o} \tag{6.1}$$

where r_i and r_o are the internal and outer radii of hollow particles, respectively. The outer radius can be measured experimentally. Radius ratio can be related to wall thickness by $t = r_o(1 - \eta)$. Average values of η and t for a batch of particles can be estimated through measurement of true particle density of HGM (ρ_{mb}) and the density of the glass (ρ_g) and using

$$\eta = \left(1 - \frac{\rho_{mb}}{\rho_g}\right)^{1/3} \tag{6.2}$$

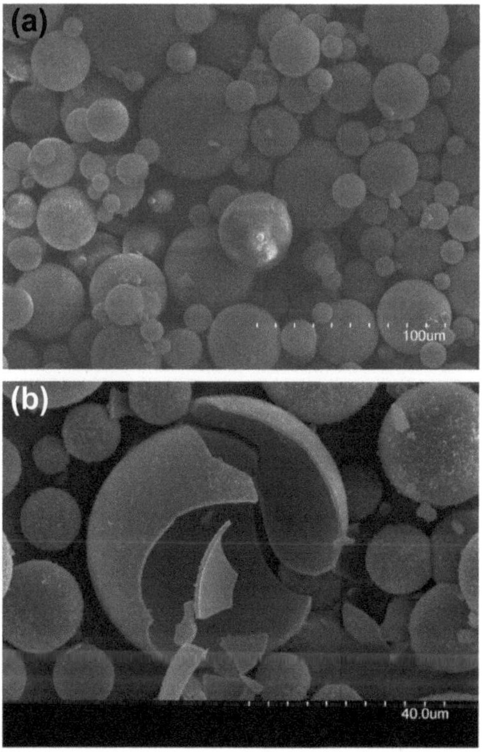

Figure 6.2 (a) A sample of Scotchlite HGMs manufactured by 3M and (b) a broken HGM of 460 kg/m^3 density showing a relatively large diameter compared to thin walls.

Table 6.1 Properties of 3M™ Glass Bubbles

Microsphere Density[a] (kg/m^3)	Average Diameter[a] (μm)	Wall Thickness (μm)	Radius Ratio (η)
220	35	0.52	0.970
320	40	0.88	0.956
380	40	1.05	0.947
460	40	1.29	0.936

[a]*Nominal values obtained from the manufacturer's datasheet.*

where ρ_{mb} and ρ_g can be experimentally measured using a pycnometer. The calculated wall thickness for four types of 3M HGM particles is given in Table 6.1

The wall thickness values for these particles vary from 520 nm to 1.29 μm. The η values for these particles are calculated to be between 0.93 and 0.98. It can be seen from Table 6.1 that although the particles do not have significant variation in outer diameter, they have different wall thicknesses, which reflects in their true particle densities. Use of particles with radius ratio lower than 0.9 is not common because such particles would be denser and limit the possibility of weight saving. The estimate of η derived from such calculations is used for developing syntactic foams with the desired density and properties.

Fabrication of Syntactic Foams

A vast majority of syntactic foams reviewed in this chapter are fabricated by mechanical mixing process. A commonly used fabrication method for epoxy/HGM syntactic foams is depicted in Figure 6.3. In the first step, HGMs are added to epoxy resin and mechanically mixed at a very slow stirring speed to minimize the damage to microspheres. In the next step, the hardener or catalyst is added and stirred. The mixture is then cast in molds and cured as per the requirements of the resin. To fabricate syntactic foams containing high volume fraction of HGMs, the resin can be heated to a higher temperature to reduce its viscosity. Another viable option is to add a suitable diluent to the resin. However, diluents should be carefully selected because they may adversely affect the mechanical properties of syntactic foams. Most epoxy manufacturers provide detailed

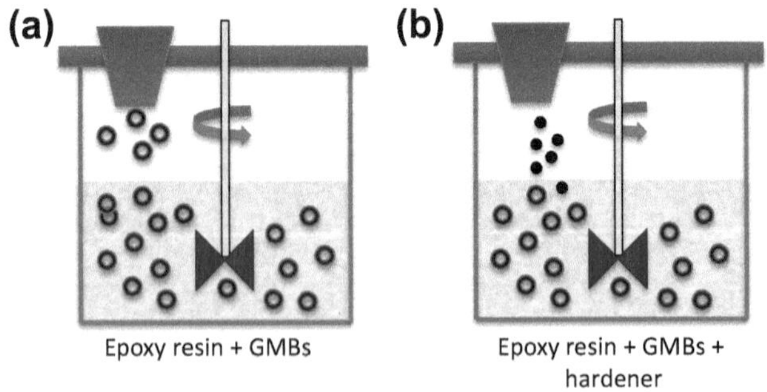

Figure 6.3 Illustration of a laboratory scale syntactic foam fabrication method.

guides on curing of the resin with different hardener choices with and without commonly used diluents.

Since HGMs have less than half the density than the matrix resin, HGMs tend to float out in the resin during curing if their volume fraction is lower than 0.3 in the syntactic foams. On the contrary, mixing over 60 vol% HGMs is difficult because they tend to break during stirring and their wetting and clustering become issues during processing. Therefore, epoxy syntactic foams with HGM volume fractions in the range 0.3–0.6 are fabricated in most studies.

Among other methods, flotation of HGMs in a column of liquid resin and infiltration of resin in a bed of particles have been used to fabricate syntactic foams. The infiltration method could be especially useful for near-net or net-shaped parts of syntactic foams.

Mechanical Properties

The properties of syntactic foams can be tailored either by changing the volume fraction of HGMs or by using HGMs of different wall thicknesses [3]. Establishing a quantitative understanding of the structure–property relationships for these materials under different loading conditions would allow tailoring these materials based on the need of the application.

Compressive Properties

Compressive stress–strain curves of epoxy matrix syntactic foams containing different volume fractions of HGMs are depicted in Figure 6.4(a). From the stress–strain curves it could be observed that, with increasing Φ_{mb}, the compressive strength decreased. It has also been noted that syntactic foams with higher density particles possess higher compressive strength. Different features of the compressive stress–strain graph of epoxy matrix syntactic foams are described in Figure 6.4(b). The graph can be divided into three regions: an initial linear elastic region, a stress-plateau region where stress remains nearly constant, and a densification region where the stress starts rising again. In the linear elastic region the material is subjected to uniform deformation and the stress value reaches a maximum, which has been defined by terms such as compressive strength, peak strength, or plastic stress. If Figure 6.4(a) and (b) are compared, it can be noticed that in the actual curves, the stress decreases to some extent after reaching a peak at the end of the elastic region. This drop in stress has been attributed to the start of HGM failure,

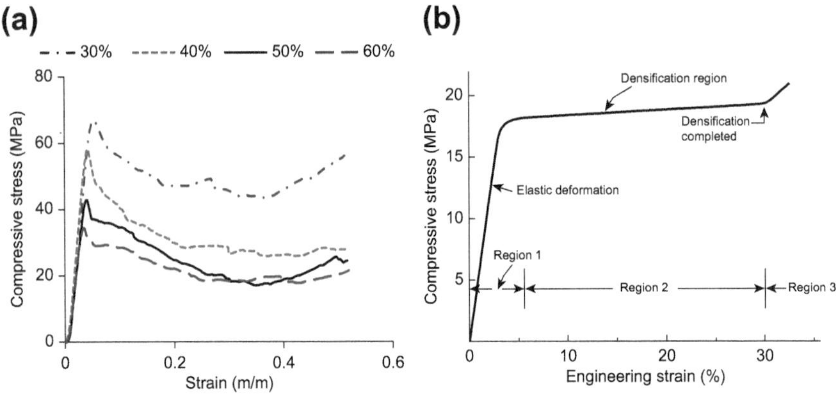

Figure 6.4 (a) Compressive stress–strain curve of epoxy matrix syntactic foam containing HGMs of 220 kg/m^3 in 30–60 vol%. (b) A generalized compressive stress–strain diagram for illustration of different features [32].

which continues throughout the remaining compression. The following region, called stress plateau, can continue up to 50–60% strain depending on the total porosity in the syntactic foams. The syntactic foams containing higher porosity content can densify more and the stress plateau can continue to higher strains. It is noted that the stress is not truly constant in the plateau region and may show a decrease, increase, or combination of both. Such characteristics depend on the relative stiffness and strength of the particle and matrix material and particle volume fraction in syntactic foam. The large stress plateau results in high-energy absorbing capability of syntactic foams. When significant amounts of HGMs are crushed and material densifies sufficiently, the stress level starts to increase again on further compression.

Crushing of HGMs is an important failure mechanism under compression. Therefore, the effects of volume fraction and wall thickness of HGMs on compressive properties of epoxy syntactic foams have been widely studied. An understanding of mechanical properties as a function of density could lead to development of structures with reduced structural weight without any penalty on mechanical properties.

The compressive strength and modulus of epoxy syntactic foams are plotted with respect to density in Figure 6.5(a) and (b), respectively. The data for these figures are obtained from Refs [3,26,30,33]. Figure 6.5(a) shows a general trend that with increasing density, the compressive strength increases. It can be seen from the figures that all the data are

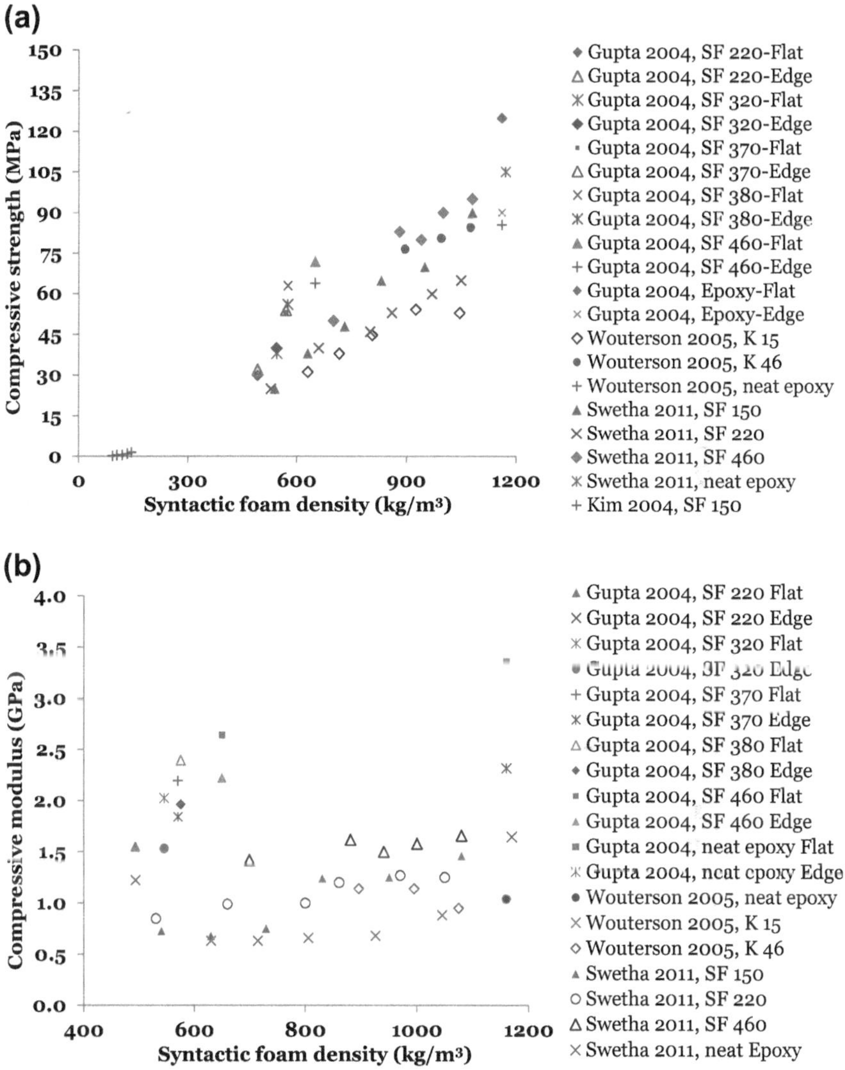

Figure 6.5 Compressive (a) strength and (b) modulus of epoxy/HGM syntactic foams with respect to the composite density. *The compressive strength data are obtained from Gupta and Woldesenbet [3], Kim and Plubrai [33], Wouterson et al. [26], Swetha and Kumar [30]. The compressive modulus data are obtained from Gupta and Woldesenbet [3], Wouterson et al. [26], Swetha and Kumar [30].*

covered within a narrow band with a slope of 0.13 MPa/kg/m^3. It does not appear that the approach of either changing the wall thickness or volume fraction has a more prominent effect on the compressive strength. It is known that some of the HGMs fracture during syntactic foam synthesis and the broken particles act like solid fillers in the resin and increase the density of the syntactic foam. In addition, mechanical mixing process of syntactic foam causes entrapment of air voids in the resin, which can be up to 5−10 vol%. Such voids result in decrease in density of syntactic foams. Despite the possibility of having defects in syntactic foams, the compressive strength does not seem to show any strong dependence on one parameter and stays on course with respect to the foam density.

The trend for compressive modulus, shown in Figure 6.5(b), is more open. If a representative density of 600 kg/m^3 is taken, the modulus is found to vary from about 0.6−2.5 GPa. The choice of particles plays an important role in this case. It is observed that the modulus of syntactic foams increases with increasing volume fraction of thick-walled particles, whereas modulus decreases with the increasing volume fraction of thin-walled particles. The ratio of modulus of the particle material to the matrix material defines the limit between the thick- and thin-walled particles for a given particle-matrix system. Extensive theoretical and simulation studies have been conducted on determining the failure mode of particles in syntactic foams. It is found that shear failure and particle crushing may occur before particle wall buckling in syntactic foams [34,35].

The graphs presented in Figure 6.5 are helpful in selecting appropriate HGMs in epoxy matrix for obtaining a predetermined set of density and mechanical properties. However, there is a need to design new methods to shift these graphs toward lower densities and higher strength values in order to increase the effectiveness of syntactic foams in various applications. It is observed in Figure 6.5 that a maximum compressive strength and modulus of 100 MPa and 2.7 GPa, respectively, are currently obtained in epoxy syntactic foams.

Tensile Properties

A representative set of tensile stress−strain curves is shown in Figure 6.6, where HGMs of two different densities, 220 and 460 kg/m^3, are used in 30−60 vol% quantities. These curves show a linear stress−strain relationship, followed by brittle failure. Epoxy matrix is the load bearing continuous phase that plays an important role in defining the

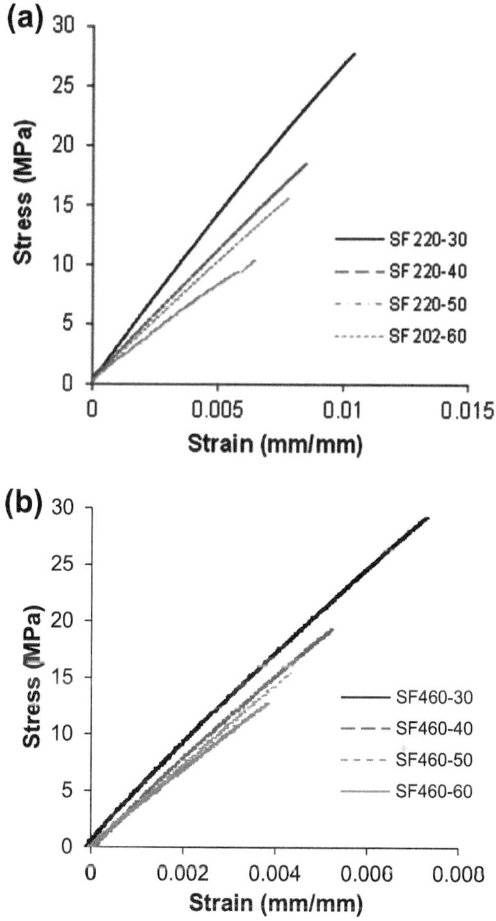

Figure 6.6 Tensile stress–strain curves of epoxy/HGM syntactic foams containing particles of (a) 220 and (b) 460 kg/m^3 in 30–60 vol%. *Adapted from Gupta and Nagorny [27].*

tensile properties and failure of epoxy/HGM syntactic foams. Epoxy resins show brittle failure mode under tensile loading, which is also reflected in the failure of the syntactic foams of these resins. The stress–strain curves under tensile loading are remarkably different from the curves obtained under compression. The elongated stress-plateau region under compression represents the exceptional energy absorption capability of syntactic foams, which is not present under tension. This is one of the main reasons that most of the syntactic foam applications rely on their compressive properties. For example, hydrostatic compression encountered in underwater applications is ideal for obtaining benefits from the properties of syntactic foams in structural applications.

The trends in tensile strength and modulus are analyzed with respect to foam density in Figure 6.7(a) and (b), respectively, for various types of epoxy/HGM syntactic foams. The trends observed in these figures are qualitatively similar to those observed for compressive strength and modulus. It is observed that the higher density foams have higher tensile strength. Considering a linear variation in tensile strength with respect to syntactic foam density in the data presented in Figure 6.7(a), the slope of the data set is calculated as 0.06 MPa/kg/m^3. This result implies that all factors affecting the syntactic foam density also affect the tensile strength in the similar way. It should also be noted that interfacial strength is not a parameter in the existing studies reported in Figure 6.7(a). Some improvement in strength may occur if the HGM–epoxy interface is made stronger to enable a more effective stress transfer.

It can also be observed that the tensile modulus of epoxy/HGM syntactic foam decreases with increasing volume fraction of thin-walled HGMs, but the trend is reversed for thick-walled HGMs. Effect of interfacial debonding on tensile and compressive response has been studied in detail in theoretical and simulation studies. It is observed that the interfacial cracks tend to close under compression and the compressing matrix is able to transfer stress to particles even when interfacial flaws or cracks are present [48,49]. However, opening of interfacial cracks or debonds under tensile loading conditions leads to degradation in properties.

The maximum values of tensile modulus and tensile strength across all compositions of epoxy/HGM syntactic foam were found to be 3.75 GPa and 45 MPa, respectively. A comparison between Figures 6.5 and 6.7 suggests that the maximum tensile strength, which is about 45 MPa, is significantly lower than the maximum compressive strength which is about 100 MPa. This large difference clearly reflects the fact that in current applications epoxy syntactic foams are normally used to obtain benefits of compressive properties. The tensile strength of epoxy syntactic foam has stronger dependence on the properties of the resin, defects such as matrix porosity and HGM–epoxy debonding, compared to compressive strength, which have a stronger dependence on the HGM properties.

Low tensile strength of syntactic foams has been a challenge in developing new applications. A possible way of overcoming this problem is to reinforce the syntactic foams with micro or nano-scale reinforcements [2]. Glass fibers, carbon fibers, carbon nanotubes, and carbon nano-fibers have been incorporated in syntactic foams to improve their tensile strength.

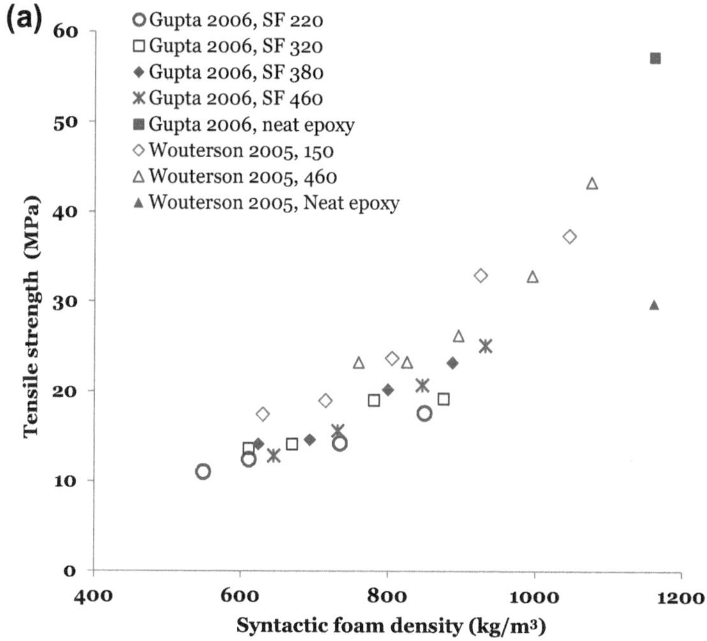

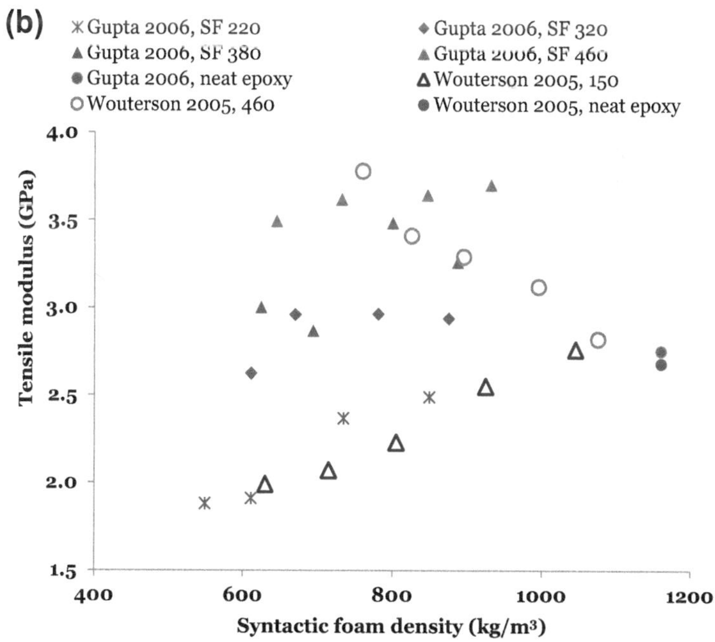

Figure 6.7 Tensile (a) strength and (b) modulus of epoxy/HGM syntactic foams with respect to the composite density. *The tensile strength data are obtained from Wouterson et al. [26], Gupta and Nagorny [27]. The tensile modulus data are obtained from Wouterson et al. [26], Gupta and Nagorny [27].*

Availability of theoretical models is useful in identifying compositions that can provide the desired set of properties in syntactic foams. One of the challenges in developing the models for syntactic foams is the high volume fraction of HGMs in such composites, where particle to particle interactions become important and simpler solutions developed for dilute dispersion of particles in the matrix are no longer applicable. Several theoretical approaches are available now [36–38]. Simulation studies are also available where from simple single particle models to complex randomly dispersed multiparticle models have been analyzed [39,40]. A model based on a differential scheme is used here to illustrate the possibility of predicting the properties of epoxy/HGM syntactic foams [38]. In this model, a homogenization scheme is used to compute the modulus of a dilute dispersion of HGMs in a matrix in the first step. In the next step, the homogenized material is taken as the matrix and a small volume fraction of particles is added again. This scheme is iteratively used until the desired volume fraction of particles is obtained. The differential scheme is given by

$$\frac{dE}{E} = f_E(E_b, \nu_b, E_m, \nu_m, \eta)\frac{d\Phi_{mb}}{(1 - \Phi_{mb}/\Phi_m)} \tag{6.3}$$

$$\frac{d\nu}{\nu} = f_\nu(E_b, \nu_b, E_m, \nu_m, \eta)\frac{d\Phi_{mb}}{(1 - \Phi_{mb}/\Phi_m)} \tag{6.4}$$

where E_b and ν_b are elastic modulus and Poisson's ratio of hollow particle material and E_m and ν_m are elastic modulus and Poisson's ratio of matrix material, respectively and Φ_{mb} is the volume fraction of the hollow filler. Detailed expressions for Eqns (6.3) and (6.4) can be found in [38]. The experimental values along with the theoretical estimates are presented in Figure 6.8. It is observed in this figure that the model predictions are in close agreement with the experimental values. Some deviations are observed between predicted and experimental values only at 60 vol% HGMs, which is close to the packing limit.

Flexural Properties

The flexural load–displacement curves for epoxy/HGM syntactic foams are similar to the tensile load–displacement curves, where load varies linearly with respect to displacement until failure [50]. The flexural behavior of syntactic foams is dominated by tensile properties of the specimen. The failure of the specimen occurs from tensile to the compressive side of the specimen. Flexural strength and modulus values

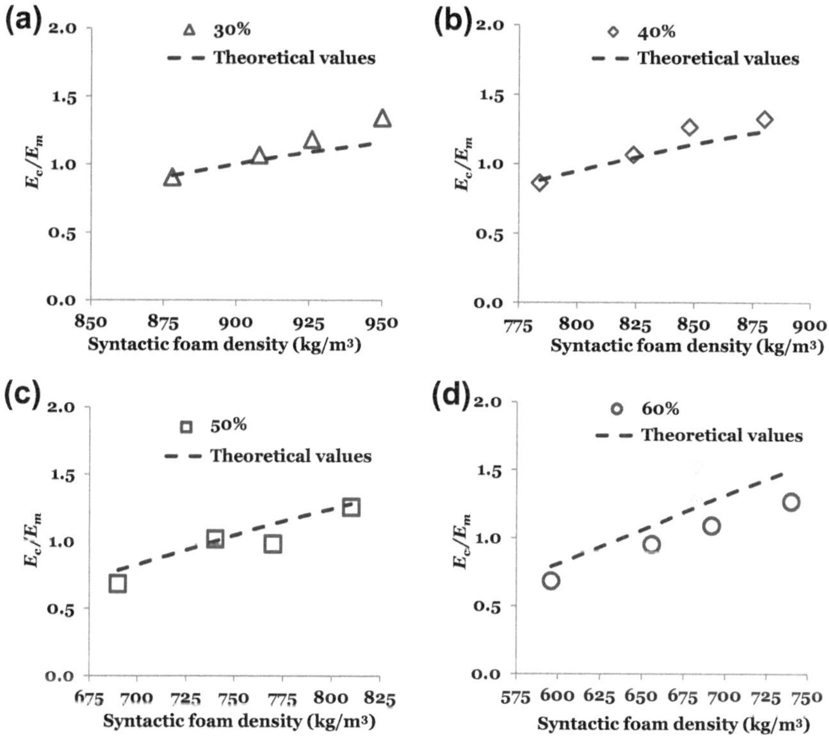

Figure 6.8 Plot of the normalized tensile modulus for epoxy/HGM syntactic foam containing Φ_{mb} of (a) 30%, (b) 40%, (c) 50%, and (d) 60%, along with the predictions from a theoretical model [38]. *The experimental data are obtained from Gupta and Nagorny [27].*

are extracted from Wouterson et al. and are plotted with respect to density in Figure 6.9 [26]. The maximum flexural strength and modulus of epoxy/HGM syntactic foams are measured to be 58 MPa and 3.8 GPa, respectively, which are close to those observed for tensile modulus of similar compositions.

The flexural strength and modulus for neat epoxy were 80 MPa and 2.8 GPa. The flexural strength of epoxy/HGM syntactic foams is lower. The presence of matrix porosity, especially on the surface, and surface defects can result in lower flexural strength and premature failure. The behavior of flexural modulus was similar to that of tensile modulus, where it can be observed that modulus decreased with increasing volume fraction of thin-walled HGMs whereas the trend is reversed for thick-walled HGMs. To the best of author's knowledge, currently there is

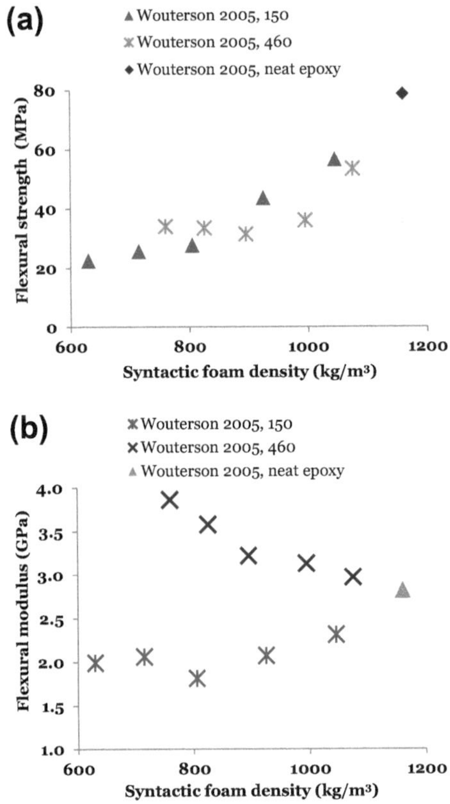

Figure 6.9 Plot of flexural (a) strength and (b) modulus of epoxy/HGM syntactic foams with respect to the composite density. *The experimental data are obtained from Wouterson et al. [26].*

only one study focused on the flexural properties of epoxy/HGM syntactic foams. A better understanding of flexural properties requires more systematic studies at various compositions of epoxy syntactic foams. In other syntactic foam systems, including vinyl ester/HGM syntactic foams, it is observed that the standard deviation in the flexural strength is larger than the compressive strength values. The flexural failure is very sensitive to surface properties such as surface roughness and presence of air voids in the matrix. Such defects can lead to premature failure and provide a large standard deviation in the strength values.

Electrical Properties

The demand for materials with low dielectric constant and low density has increased in electronic components due to the miniaturization of circuit boards used in electronic devices. Polymers such as epoxy resins have been used in electrical and electronic fields as insulators, dielectrics and as underfills in circuit boards [41,42]. However, they have low strength and stiffness, which is a limiting factor in making large size boards. Porosity contributes to decreasing the dielectric constant in syntactic foams because air has a low dielectric constant of 1. The porosity fraction in the syntactic foam can be altered by means of HGM volume fraction and wall thickness. Table 6.2 lists the studies that are focused on understanding the influence of volume fraction of HGMs on the dielectric constant at different frequencies. Several available studies on measuring dielectric constant are related to studying environmental degradation of syntactic foams and using dielectric constant as a measure of polymer degradation.

It can be noted from studies presented in Table 6.2 that dielectric constant decreases with increase in volume fraction of HGMs. The dielectric loss, which defines the amount of energy loss as heat is lower in epoxy syntactic foams compared to neat epoxy. Decrease in wall thickness also increases the air void size in HGMs. However, in the most widely used particles, the wall thickness changes over a small range, 0.5−1.3 μm for HGM diameter of 40 μm. Therefore, the effect of wall thickness change on the dielectric constant of syntactic foam is usually not well resolved. However, wall thickness has a strong effect on other properties such as elastic modulus and a combination of wall thickness and volume fraction can be used for tailoring two or more properties simultaneously. Theoretical models can be very useful in understanding the trends that exist on the effect of HGM wall thickness and volume fraction on the dielectric constant of syntactic foams.

In a recent study, the dielectric constant of vinyl ester/HGM syntactic foams has been predicted based on the dielectric constant of the matrix and the HGM [21]. The dielectric constant of HGM is obtained from the dielectric constant of glass and the geometric parameter η using the theory of Maxwell by

$$\varepsilon_{mb} = \left[\frac{1 - 2\eta^3 \frac{(\varepsilon_g - 1)^2}{(\varepsilon_g + 2)(2\varepsilon_g + 1)} + 2(1 - \eta^3)\frac{(\varepsilon_g - 1)}{(\varepsilon_g + 2)}}{1 - 2\eta^3 \frac{(\varepsilon_g - 1)^2}{(\varepsilon_g + 2)(2\varepsilon_g + 1)} - (1 - \eta^3)\frac{(\varepsilon_g - 1)}{(\varepsilon_g + 2)}} \right] \tag{6.5}$$

Table 6.2 Studies on Dielectric Constant of Epoxy/HGM Syntactic Foams

References	Φ_{mb}	Test Conditions[a]	Study Outcome
[43]	0.3–0.65	$\omega = 1–100$ kHz $T = 40–120$ °C	1. The Dielectric constant decreases with increasing Φ_{mb}. 2. The Dielectric constant increases 5–10% with increase in the temperature. 3. The Impedance decreased with increasing frequency and the phase angle remained close to −90°, suggesting capacitive behavior.
[45]	0–0.51	$\omega = 1$ kHz–1 MHz	1. At 1 MHz, the dielectric constant decreased from 3.98 to 2.84 as Φ_{mb} increased from 0 to 0.51. 2. A Maximum decrease of 44% is observed in the dielectric loss at $\Phi_{mb} = 0.51$ compared to neat resin.
[44]	0.1–0.6	$\omega = 1$ MHz $T = 10$ °C	1. A Maximum of 51% decrease was observed in the dielectric constant at $\Phi_{mb} = 0.60$ compared to neat resin. 2. A maximum of 54% decrease was observed in the dielectric loss at $\Phi_{mb} = 0.6$ compared to neat resin.

[a] *ω = frequency, T = temperature.*

where ε_g and ε_a are taken as 5.6 and 1, respectively [44]. The dielectric constant can be used in conjunction with Jayasundere–Smith (J–S) equation to obtain estimates for the dielectric constant of the syntactic foams. The modified J–S equation for prediction of dielectric constant of syntactic foams is given by [21].

$$\varepsilon = \frac{\Phi_m \varepsilon_m + \Phi_{mb} \varepsilon_{mb} \left(\frac{3\varepsilon_m}{\varepsilon_{mb} + 2\varepsilon_m} \right) \left(1 + \frac{3\Phi_{mb}(\varepsilon_{mb} - \varepsilon_m)}{\varepsilon_{mb} + 2\varepsilon_m} \right)}{\Phi_m + \Phi_{mb} \left(\frac{3\varepsilon_m}{\varepsilon_{mb} + 2\varepsilon_m} \right) \left(1 + \frac{3\Phi_{mb}(\varepsilon_{mb} - \varepsilon_m)}{\varepsilon_{mb} + 2\varepsilon_m} \right)} \tag{6.6}$$

The dielectric constants of epoxy/HGM matrix syntactic foams predicted from Eqn (6.6) are presented in Figure 6.10 with experimental results obtained from Yung et al. [45]. The predicted values show a close match with the experimental values, suggesting the dielectric constant syntactic foams can be tailored. The same model has been extensively tested with vinyl ester/HGM syntactic foams and the predictions are found to be close match with experimental data [21].

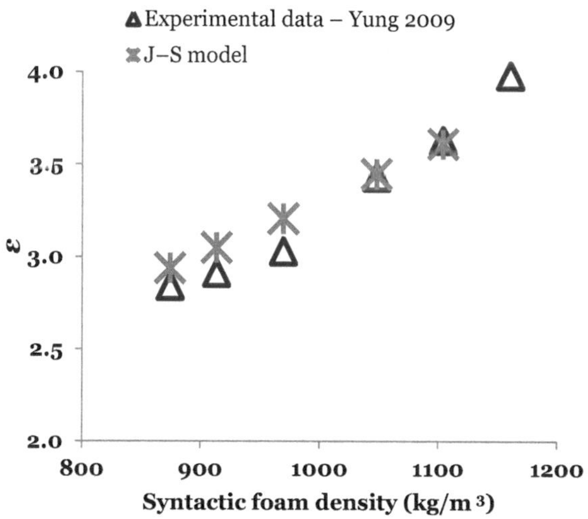

Figure 6.10 Comparison of theoretical and experimental values for the dielectric constant of epoxy/HGM syntactic foam at a frequency of 1 MHz. *The experimental data are taken from Yung et al. [45].*

Thermal Properties

Understanding CTE of syntactic foams is important for their applications in deep-sea oil pipelines and aircraft structures. Figure 6.11 depicts CTE as a function of syntactic foam density from two available studies. It can be observed that epoxy/HGM syntactic foams have lower CTE compared to that of neat epoxy matrix. This result suggests that syntactic foams possess better dimensional stability compared to neat epoxy matrix. It can also be observed that thin-walled HGMs provide better dimensional stability compared to that of thick-walled HGMs. Epoxy syntactic foams result in the least CTE value of 28.72 μm/m °C in the available studies. This value of the CTE is 54.3% lower than the CTE value of the neat epoxy resin. In another recent study, it has been shown that by simultaneously varying η and Φ_{mb} of the HGM, the CTE of the syntactic foam can be tailored [20].

The CTE of epoxy/HGM syntactic foams can be predicted using Turner's model modified to account for the wall thickness of the hollow particle, by replacing HGM with an equivalent elastic sphere [46]. The modified Turner's model is given by [20]

$$\alpha = \frac{\alpha_m \Phi_m E_m \left[(1 - 2\nu_g) + \left(\frac{1+\nu_g}{2} \right) \eta^3 \right] + \alpha_b \Phi_b E_b (1 - \eta^3)(1 - 2\nu_m)}{\Phi_m E_m \left[(1 - 2\nu_g) + \left(\frac{1+\nu_g}{2} \right) \eta^3 \right] + \Phi_b E_b (1 - \eta^3)(1 - 2\nu_m)} \tag{6.7}$$

where ν_g and ν_m are the Poisson's ratio of the HGM and matrix material, E_m and E_b are the modulus of the matrix and the HGM material. The density of the HGM used in Yung et al. [45] and Saha et al. [47] are 600 and 150 kg/m^3. The value of η was evaluated using Eqn (6.2), with the density of the glass as 2540 kg/m^3. The estimates from the modified Turner's model along with the experimental values for the epoxy/HGM syntactic foams are shown in Figure 6.12. The theoretical estimates are in close match with the experimental values and the modified Turner's model can be utilized to identify compositions of syntactic foams with the desired set of properties.

Multifunctional Syntactic Foams

It has been demonstrated in previous sections that mechanical, electrical, and thermal properties of syntactic foams can be tailored based on η

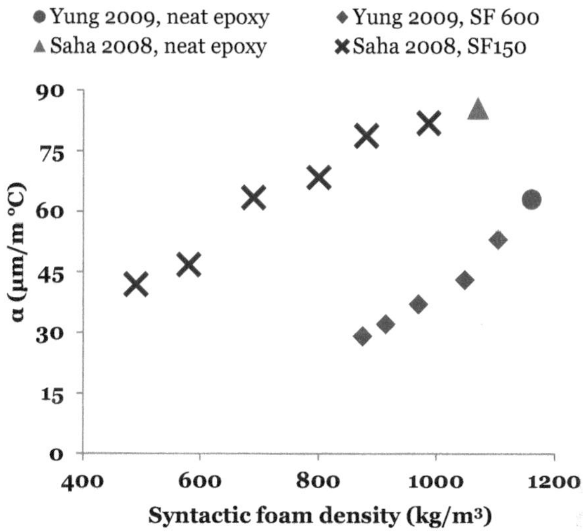

Figure 6.11 Variation of coefficient of thermal expansion (α) of epoxy/HGM syntactic foams with respect to the composite density. *The CTE data are taken from Saha et al. [47], Yung et al. [45].*

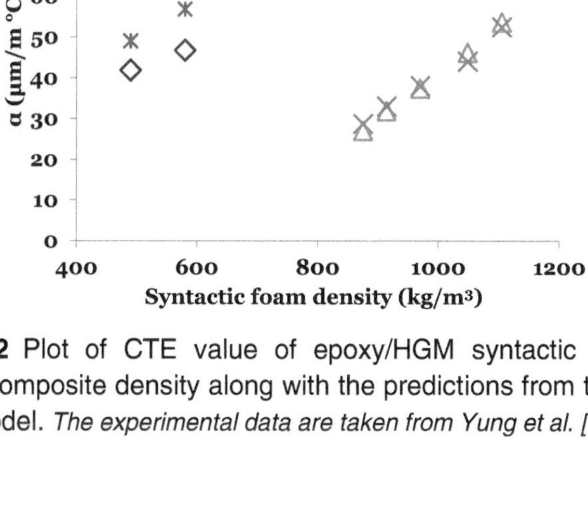

Figure 6.12 Plot of CTE value of epoxy/HGM syntactic foams with respect to composite density along with the predictions from the modified Turner's model. *The experimental data are taken from Yung et al. [45] and Saha et al. [47].*

and Φ_{mb} of HGMs. It is often desired, while designing industrial components, to have materials with two or more properties within specified range. For example, the syntactic foam used in the deckhouse of USS Zumwalt would require certain level of modulus and strength to enable construction of a structural component while it is also desired to have tailored electromagnetic properties for stealth. The concept of tailorability can be extended beyond just one property in case of syntactic foams, to simultaneously tailor two or more properties.

Experimentally validated theoretical models are used in Figure 6.13 to plot the variation of dielectric constant, CTE, density, and modulus of the

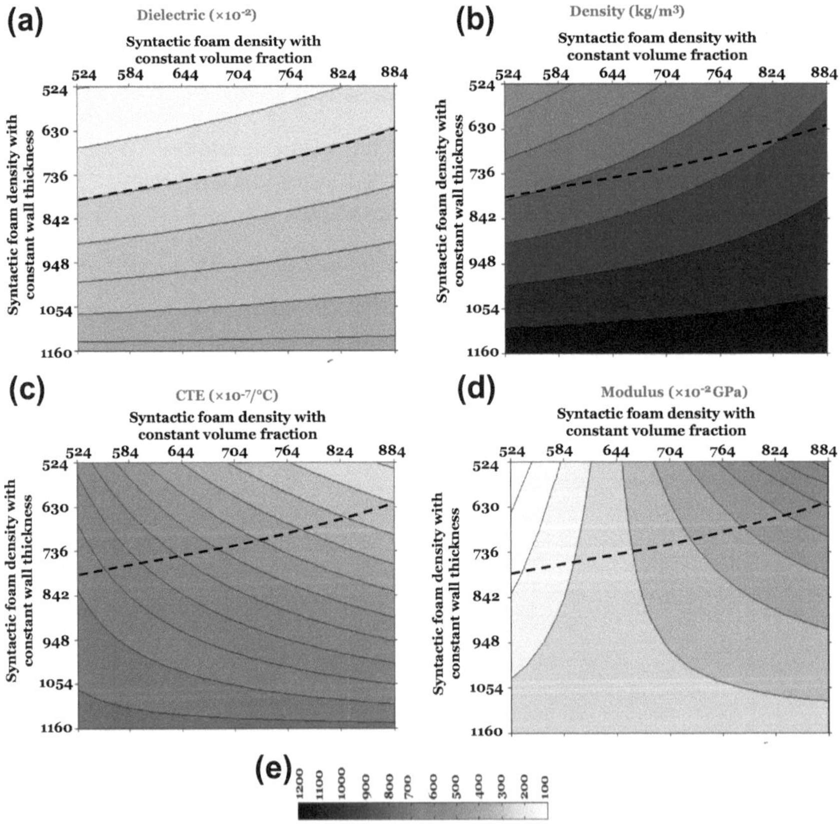

Figure 6.13 Variation of (a) dielectric constant, (b) density (kg/m^3), (c) CTE (μm/m °C), and (d) modulus (GPa) of syntactic foams with respect to syntactic foam density. The HGM density varies from 100 to 700 kg/m^3 and Φ_{mb} between 0 and 0.6. The dashed line represents a dielectric constant of 3. The scale bar in (e) represents the CTE, dielectric constant, density, and the modulus in their corresponding units.

syntactic foams with respect to the Φ_{mb} and η. The dashed line in Figure 6.13(a) represents compositions of syntactic foams with a dielectric constant of 3. This value of dielectric constant is arbitrarily selected to demonstrate the concept of tailorability. This line can be projected on to the CTE, density, and the modulus plots. From this figure it can be observed that for syntactic foams having dielectric constant of 3, the CTE can vary between 30 and 69 μm/m °C, the density from 800 to 925 kg/m^3, and the modulus values between 1.9 and 5.1 GPa. Depending on the secondary requirements of the component, appropriate composition can be selected. Solid particle-reinforced composites do not present such possibilities because they only have one parameter, volume fraction, to vary in order to change the properties of the composite.

Summary

This chapter summarizes the effects of HGMs on the mechanical, electrical, and thermal properties of epoxy matrix syntactic foams. The data for various mechanical properties such as compressive, tensile, and the flexural strength and modulus; dielectric constant and CTE have been plotted as a function of syntactic foam density. It can be identified that the two critical parameters, wall thickness and volume fraction of HGMs can be independently used to simultaneously tune multiple properties of syntactic foams. This method of controlling the two parameters of hollow particles to meet the demands for multifunctional structures shows that syntactic foams are advantageous compared to solid particulate composites.

Theoretical models to estimate tensile modulus, CTE and dielectric constant of epoxy syntactic foam are also presented in this chapter. These models are validated with experimental results. However, there is still a need for systematic studies to understand the strain rate effects on mechanical properties and temperature effects on thermal and mechanical properties. The particle to particle interaction effects can also be significant in syntactic foams and can lead to under- or overestimation of properties depending on the factors such as relative stiffness, size, and volume faction of constituents. Studies related to environmental degradation of epoxy/HGM are available only under certain conditions. Detailed studies on the residual mechanical properties, long-term exposure, and combined effect of high temperature, moisture, and ultraviolet radiation are desired for existing applications.

List of Symbols

η	Radius ratio
r_i	Inner radius of glass microsphere
r_o	Outer radius of glass microsphere
t	Wall thickness of glass microsphere
Φ_{mb}	Volume fraction of glass microsphere
Φ_m	Volume fraction of matrix material
α	Coefficient of thermal expansion of syntactic foam
α_m	Coefficient of thermal expansion of matrix material
α_b	Coefficient of thermal expansion of hollow particle material
ρ_{mb}	True particle density of glass microsphere
ρ_g	Density of glass
E_b	Elastic modulus of hollow particle material
ν_b	Poisson's ratio of hollow particle material
E_m	Elastic modulus of matrix material
ν_m	Poisson's ratio of matrix material
ε	Dielectric constant of syntactic foam
ε_g	Dielectric constant of glass
ε_a	Dielectric constant of air
ε_m	Dielectric constant of matrix material
ε_{mb}	Dielectric constant of glass microsphere

Acknowledgments

This work was funded by the Office of Naval Research grant N00014-10-1-0988. The authors thank Steven E. Zeltmann for help in manuscript preparation. The views and conclusions contained in this publication are those of the authors and should not be interpreted as presenting the official policies or position, either expressed or implied, of the ONR or the U.S. Government unless so designated by other authorized documents. Citation of manufacturers' or trade names does not constitute an official endorsement or approval of the use notwithstanding any copyright notation hereon.

References

[1] F. Shutov, Syntactic Polymer Foams. Advances in Polymer Science, Springer, New York, 1986, pp. 63–123.

[2] N. Gupta, D. Pinisetty, V.C. Shunmugasamy, Reinforced Polymer Matrix Syntactic Foams Effect of Nano and Micro-scale Reinforcement, Springer Briefs in Materials, New York, 2013.

[3] N. Gupta, E. Woldesenbet, P. Mensah, Compression properties of syntactic foams: Effect of cenosphere radius ratio and specimen aspect ratio, Composites Part A 35 (1) (2004) 103–111.
[4] F. Grosjean, N. Bouchonneau, D. Choqueuse, V. Sauvant-Moynot, Comprehensive analyses of syntactic foam behaviour in deepwater environment, J. Mater. Sci. 44 (6) (2009) 1462–1468.
[5] E.C. Hobaica, S.D. Cook, The characteristics of syntactic foams used for buoyancy, J. Cell. Plast. 4 (4) (1968) 143–148.
[6] N. Gupta, E. Woldesenbet, Kishore, S. Sankaran, Response of syntactic foam core sandwich structured composites to three-point bending, J. Sandwich Struct. Mater. 4 (3) (2002) 249–272.
[7] G. Tagliavia, M. Porfiri, N. Gupta, Vinyl ester—glass hollow particle composites: dynamic mechanical properties at high inclusion volume fraction, J. Compos. Mater. 43 (5) (2009) 561–582.
[8] Z. Chen, Z. Huang, Y. Qin, M. Shi, Q. Mei, M. Zhang, Effect of glass microsphere size on compressive strength of syntactic foams, Adv. Mater. Res. 321 (2011) 7–10.
[9] K. Carlisle, M. Koopman, K. Chawla, R. Kulkarni, G. Gladysz, M. Lewis, Microstructure and compressive properties of carbon microspheres, J. Mater. Sci. 41 (13) (2006) 3987–3997.
[10] L. Zhang, J. Ma, Effect of coupling agent on mechanical properties of hollow carbon microsphere/phenolic resin syntactic foam, Compos. Sci. Technol. 70 (8) (2010) 1265–1271.
[11] M. Labella, S.E. Zeltmann, V.C. Shunmugasamy, N. Gupta, P.K. Rohatgi, Mechanical and thermal properties of fly ash/vinyl ester syntactic foams, Fuel 121 (2014) 240–249.
[12] L. Licitra, D.D. Luong, O.M. Strbik III, N. Gupta, Dynamic properties of alumina hollow particle filled aluminum alloy A356 matrix syntactic foams, Mater. Des Part B 66 (2015) 504–515.
[13] M. Labella, V. Chakravarthy Shunmugasamy, O.M. Strbik, N. Gupta, Compressive and thermal characterization of syntactic foams containing hollow silicon carbide particles with porous shell, J. Appl. Polym. Sci. 131 (2014) 40689.
[14] W. Xie, H. Yan, Q. Mei, M. Du, Z. Huang, Compressive and fracture properties of syntactic foam filled with hollow plastic bead (HPC), J. Wuhan Univ. Technol., Mater. Sci. Ed. 22 (3) (2007) 499–501.
[15] S.S. Samsudin, Z.M. Ariff, Z. Zakaria, A.A. Bakar, Development and characterization of epoxy syntactic foam filled with epoxy hollow spheres, eXPRESS Polym. Lett. 5 (7) (2011) 653–660.

[16] L. Yusriah, M. Mariatti, Effect of hybrid phenolic hollow microsphere and silica-filled vinyl ester composites, J. Compos. Mater. 47 (2) (2013) 169–182.
[17] L. Wang, X. Yang, J. Zhang, C. Zhang, L. He, The compressive properties of expandable microspheres/epoxy foams, Composites, Part B Eng. 56 (2014) 724–732.
[18] G. Hu, D. Yu, Tensile, thermal and dynamic mechanical properties of hollow polymer particle-filled epoxy syntactic foam, Mater. Sci. Eng: A 528 (15) (2011) 5177–5183.
[19] V. Shunmugasamy, D. Pinisetty, N. Gupta, Thermal expansion behavior of hollow glass particle/vinyl ester composites, J. Mater. Sci. 47 (14) (2012) 5596–5604.
[20] V. Shunmugasamy, D. Pinisetty, N. Gupta, Viscoelastic properties of hollow glass particle filled vinyl ester matrix syntactic foams: effect of temperature and loading frequency, J. Mater. Sci. 48 (4) (2012) 1685–1701.
[21] V. Shunmugasamy, D. Pinisetty, N. Gupta, Electrical properties of hollow glass particle filled vinyl ester matrix syntactic foams, J. Mater. Sci. 49 (1) (2014) 180–190.
[22] B. John, C.P.R. Nair, K.N. Ninan, Effect of nanoclay on the mechanical, dynamic mechanical and thermal properties of cyanate ester syntactic foams, Mater. Sci. Eng: A 527 (21–22) (2010) 5435–5443.
[23] K. Okuno, R.T. Woodhams, Mechanical properties and characterization of phenolic resin syntactic foams, J. Cell. Plast. 10 (1974) 237–244.
[24] G. Gladysz, B. Perry, G. McEachen, J. Lula, Three-phase syntactic foams: structure-property relationships, J. Mater. Sci. 41 (13) (2006) 4085–4092.
[25] B. Song, W. Chen, D.J. Frew, Dynamic compressive response and failure behavior of an epoxy syntactic foam, J. Compos. Mater. 38 (11) (2004) 915–936.
[26] E.M. Wouterson, F.Y.C. Boey, X. Hu, S.-C. Wong, Specific properties and fracture toughness of syntactic foam: effect of foam microstructures, Compos. Sci. Technol. 65 (11–12) (2005) 1840–1850.
[27] N. Gupta, R. Nagorny, Tensile properties of glass microsphere-epoxy resin syntactic foams, J. Appl. Polym. Sci. 102 (2) (2006) 1254–1261.
[28] B. Song, W.W. Chen, W.Y. Lu, Mechanical characterization at intermediate strain rates for rate effects on an epoxy syntactic foam, Int. J. Mech. Sci. 49 (12) (2007) 1336–1343.

[29] P. Li, N. Petrinic, C.R. Siviour, R. Froud, J.M. Reed, Strain rate dependent compressive properties of glass microsphere epoxy syntactic foams, Mater. Sci. Eng: A 515 (1−2) (2009) 19−25.
[30] C. Swetha, R. Kumar, Quasi-static uni-axial compression behaviour of hollow glass microspheres/epoxy based syntactic foams, Mater. Des. 32 (8−9) (2011) 4152−4163.
[31] N. Gupta, S. Zeltmann, V. Shunmugasamy, D. Pinisetty, Applications of polymer matrix syntactic foams, JOM: J. Miner., Met., Mater. Soc. 66 (2) (2014) 245−254.
[32] N. Gupta, K. Kishore, E. Woldesenbet, S. Sankaran, Studies on compressive failure features in syntactic foam material, J. Mater. Sci. 36 (18) (2001) 4485−4491.
[33] H.S. Kim, P. Plubrai, Manufacturing and failure mechanisms of syntactic foam under compression, Composites Part A: Appl. Sci. Manuf. 35 (9) (2004) 1009−1015.
[34] A. Shams, M. Porfiri, Axisymmetric static and dynamic buckling of hollow microspheres, Int. J. Non-Linear Mech. 61 (0) (2014) 19−31.
[35] A. Shams, M. Aureli, M. Porfiri, Nonlinear buckling of a spherical shell embedded in an elastic medium with imperfect interface, Int. J. Solids Struct. 50 (14−15) (2013) 2310−2327.
[36] L. Bardella, F. Genna, On the elastic behavior of syntactic foams, Int. J. Solids Struct. 38 (40−41) (2001) 7235−7260.
[37] P.R. Marur, Effective elastic moduli of syntactic foams, Mater. Lett. 59 (14−15) (2005) 1954−1957.
[38] M. Porfiri, N. Gupta, Effect of volume fraction and wall thickness on the elastic properties of hollow particle filled composites, Compos. Part B: Eng. 40 (2) (2009) 166−173.
[39] L. Bardella, A. Sfreddo, C. Ventura, M. Porfiri, N. Gupta, A critical evaluation of micromechanical models for syntactic foams on the basis of three-dimensional finite element analyses, Mech. Mater. 50 (2012) 53−69.
[40] L. Bardella, F. Malanca, P. Ponzo, A. Panteghini, M. Porfiri, A micromechanical model for quasi-brittle compressive failure of glass-microspheres/thermoset-matrix syntactic foams, J. Eur. Ceram. Soc. 34 (11) (2014) 2605−2616.
[41] K.J. Miller, K.N. Collier, H.B. Soll-Morris, R. Swaminathan, M.E. McHenry, Induction heating of FeCo nanoparticles for rapid rf curing of epoxy composites, J. Appl. Phys. 105 (7) (2009), 07E714-711-713.

[42] P.A. Kohl, Low-dielectric constant insulators for future integrated circuits and packages, Annu. Rev. Chem. Biomol. Eng. 2 (2011) 379–401.

[43] N. Gupta, S. Priya, R. Islam, W. Ricci, Characterization of mechanical and electrical properties of epoxy-glass microsphere syntactic composites, Ferroelectrics 345 (1) (2006) 1–12.

[44] B. Zhu, J. Ma, J. Wang, J. Wu, D. Peng, Thermal, dielectric and compressive properties of hollow glass microsphere filled epoxy-matrix composites, J. Reinf. Plast. Compos. 31 (19) (2012) 1311–1326.

[45] K.C. Yung, B.L. Zhu, T.M. Yue, C.S. Xie, Preparation and properties of hollow glass microsphere-filled epoxy-matrix composites, Compos. Sci. Technol. 69 (2) (2009) 260–264.

[46] J. Nji, G. Li, A CaO enhanced rubberized syntactic foam, Compos. Part A: Appl. Sci. Manuf. 39 (9) (2008) 1404–1411.

[47] M.C. Saha, S. Nilufar, M. Major, S. Jeelani, Processing and performance evaluation of hollow microspheres filled epoxy composites, Polym. Compos. 29 (3) (2008) 293–301.

[48] G. Tagliavia, M. Porfiri, N. Gupta, Analysis of hollow inclusion-matrix debonding in particulate composites, Int. J. Solids Struct. 47 (16) (2010) 2164–2177.

[49] G. Tagliavia, M. Porfiri, N. Gupta, Elastic interaction of interfacial spherical-cap cracks in hollow particle filled composites, Int. J. Solids Struct. 48 (7-8) (2011) 141–1153.

[50] R. Maharsia, N. Gupta, H.D. Jerro, Investigation of flexural strength properties of rubber and nanoclay reinforced hybrid syntactic foams, Mater. Sci. Eng. A 417(1-2) (2006) 249–258.

7 Hollow Glass Microspheres in Polyurethanes

Baris Yalcin

Polyurethane Basics

Polyurethanes (PUs) (or urethane polymers) are one of the most versatile materials used today for numerous applications ranging from flexible foam in upholstered furniture to rigid foam as insulation in walls, roofs, and pipes; to thermoplastics used in medical devices and footwear; to coatings, adhesives, sealants, and elastomers used on floors and automotive interiors.

PUs can be made available in multitude of forms (liquid, foam, or solid) and properties (eraser-soft to bowling-ball-hard). PUs are composed of two principal raw materials: isocyanates and polyols, brought together with catalysts and chain extenders (cross-linkers) and a variety of additives including blowing agents, fillers, flame retardants, antioxidants, plasticizers, and pigments.

PUs are in a class of compounds called reaction polymers, which include epoxies, unsaturated polyesters, and phenolics. The basis of PU chemistry is the high reactivity of isocyanates. Figure 7.1 shows the classical two-step PU synthesis, where a diisocyanate and a polyhydroxyl (polyol) first react to form a prepolymer which in turn is extended by a low molecular weight hydroxyl compound (diol) into a long chain segmented linear PU. An example phase-separated linear morphology such as that shown below is formed by the flexible phase (product of long chain polyol and isocyanate) and a rigid phase (product of isocyanate and chain extender). The phase-separated morphology can change depending on the molecular weight and functionality of the isocyanate, polyols, and chain extenders. Thermoset PUs formed by polyol or chain extenders with hydroxyl or amine functionalities more than 2, still exhibit phase-separated soft and hard domains but not in linear form as seen in thermoplastic urethanes.

In practice, the broad spectrum of properties is achieved by the appropriate selection of polyols and additives. The number of isocyanates used is limited to a few types of aromatic tolylenediisocyanate (TDI), 4,4′-methylenediphenyl diisocyanate (MDI), custom-formulated isocyanate components, and in special cases aliphatic isocyanates (e.g., HDI,

Hollow Glass Microspheres for Plastics, Elastomers, and Adhesives Compounds
http://dx.doi.org/10.1016/B978-1-4557-7443-2.00007-4

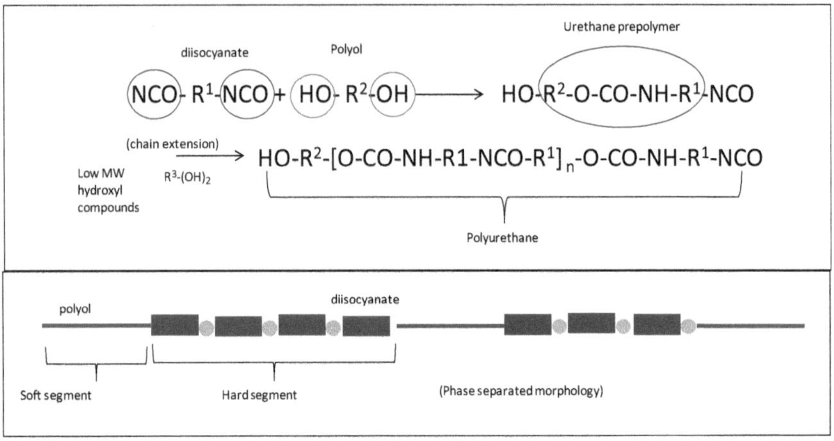

Figure 7.1 Two-step polyurethane reaction.

IPDI) where special effects such as light stability and optical clarity are needed. Aromatic isocyanates are more economical than aliphatic isocyanates and they have higher reactivity. Isocyanates are characterized by the percentage of NCO content and their functionality, which indicates how many NCO groups a molecule contains. Thermal stability of PU is strongly dependent on the presence of hydrogen bonding and the isocyanate type used. Hydrogen bonding dissociates at around 80 °C. Thermal stability of different isocyanates is as follows: HMDI > MDI > HTDI > TDI > HDI.

More subtle changes can be made to the properties of PU by varying the isocyanate/hydroxyl ratio, adding cross-linking agents, using a different catalyst, and changing the polymerization process. By changing these variables, PU polymers can be produced which overlap the hardness of soft rubbers to hard thermoplastics.

PU properties depend greatly on the chemistry (ether- or ester-based), functionality (number of hydroxyl groups), and molecular weight of the polyols and chain extenders (cross-linkers) used. Thermoplastic polyurethanes (TPUs) based on ester polyols have enhanced oil and chemical resistance and provide excellent abrasion resistance whereas the polyether TPUs exhibit better low temperature flexibility and durable against microbial attack and provide hydrolysis resistance.

The number of functional groups of the polyols, chain extenders, and isocyanates used in the PU synthesis determines the nature of the formed cross-links which could be heat reversible (thermoplastic) or irreversible (thermoset). When these raw materials are limited to bifunctional chemistries, for example, 4,4-diphenylmethane diisocyanate (MDI) with long chain diols and 1,4-butanediol as a chain extender, thermoplastic linear PUs are formed. The thermoplastic processability of this material is possible due to the structure of TPU, which consists essentially of linear macromolecules.

Thermoset PUs differ from the thermoplastics in that irreversible cross-links are formed when the polymers are chemically cured. Cross-linking takes place both chemically (triols, polyfunctional isocyanates) and physically by phase separation into generally amorphous urea rigid segments and flexible polyol segments. The process of PU cross-linking in thermoset form can occur in many different ways:

1. In soft segments by using polyols with functionality higher than 2.
2. In hard segments by using a chain extender with functionality higher than 2 (cross-linkers).
3. In hard segments by using an excess of isocyanates and the proper catalyst, which leads to the formation of allophanate or biuret bonds that cause branching or cross-linking at a particular temperature.
4. In hard segments by using multifunctional isocyanates.
5. In hard segments by using products of isocyanate dimerization or trimerization with proper catalysts.
6. By the sulfur vulcanization of unsaturated millable PUs or cross-linking by the radical copolymerization of unsaturated PUs.

For the synthesis of cross-linked (thermoset) PUs, typically used polyols are those having a molecular weight in the range of 300−10,000 Da and functionality in the range $2 < f \leq 8$ (f—polyol functionality determined as the number of OH groups per 1 mol of compound).

Polyols of high functionality combined with multifunctional poly-isocyanates are used mainly for the synthesis of rigid PU foams. See the summary in Table 7.1.

For further thermoset PU cross-linking reactions, such as by using a chain extender with functionality higher than 2 or by using an excess of isocyanates and the proper catalyst, readers are referred to Ref. [1].

PUs can be formed in a one- or two-step prepolymer process but for thermoset PUs, the prepolymer method, where the polyol and diisocyanate are first co-reacted and then extended with a diamine or glycol, is generally preferred for thermosets since it provides better control of

Table 7.1 Polyol Functionality and Polyurethane (PU) Application

MW	Polyol Functionality Hydroxyl Groups/mol	Application
2000–10,000	2–3	Flexible PU foam
150–1000	3–8	Rigid PUs

chemical reactivity and assures that random polymer chains are established prior to cross-linking.

Hollow glass microspheres (HGMs) have been used in thermoplastic, thermoset, and foamed PUs.

HGMs in Thermoplastic PUs

HGMs can be incorporated into thermoplastic urethanes by standard melt compounding or reactive extrusion route.

Wilson et al. showed in the 1990s that reactive extrusion is beneficial for compounding HGMs with low isostatic crush strength, that is, 10,000 psi, into TPUs with high survival rates [2]. It was discovered that adding the spheres during the in situ polymerization of the resin greatly reduces or eliminates the fracturing of the spheres. In addition, they have shown that large volume loadings of HGMs could be realized using reactive extrusion which is useful as a masterbatch for compounding into other resins, particularly resins filled with fiberglass reinforcement. Significant reduction or elimination of HGM fracture was believed to be due to the masterbatch resin covering the glass spheres which protects them from breakage by lessening the effects of the shear forces exerted by the mechanical action of the compounding apparatus and the abrasive contact with the glass fibers.

The in situ polymerization process is a reactive extrusion process where the polymer reactants are added to the reactor portion of the apparatus where the polymerization begins. The reacting mixture is then passed into the extruder portion of the apparatus during which time the polymerization reaction is completed before the polymer resin exits through the extruder die. The hollow glass spheres are added after about 50–90% of the material is polymerized. For example, to make an in situ-polymerized thermoplastic PU masterbatch containing hollow glass spheres, the two resin components, that is, a blend of polyol and diol and MDI, are heated to about 100 °C, separately. The blend and the MDI are then mixed together in the reactor and a standard high temperature random melt polymerization reaction begins. Due to the exothermic nature of the reaction, the temperature increases to about 200−280 °C in about 1−5 min along with an increase in the viscosity of the material. After 90% of the reaction is completed, the material is passed into the extruder, where it is mixed under high shear forces and polymerization is completed before the polymer resin exits through the extruder die. The hollow glass spheres are added to the polymerizing resin just before it enters into the extruder. It was found that a higher molecular weight polymer also aided in diminishing the fracture of the glass spheres. When the ratio of isocyanate (NCO) to hydroxyl (OH) is about 1:1, then a polymer having a molecular weight of at least about 200,000 was advantageous.

Table 7.2 compares the results of the percent damaged spheres (0.6 g/cc 10,000 Psi) in TPU-HGM masterbatch resin prepared by traditional melt compounding versus reactive in situ extrusion method [2]. Table 7.2 also shows how HGMs in these masterbatch resins survive when melt compounded and injection molded in a subsequent step using with a glass fiber compound. The masterbatch resin and unfilled compounding resin were TPUs comprising MDI, a polyol having a molecular weight between 3000 and 4000, adipic acid, and 1,4-butane diol. All specific gravity measurements were made on injection-molded plaques. Table 7.2 shows that in situ reactive compounding results in high survival rates for HGMs (17% vs 3−5%). Furthermore, when the masterbatches are used further with glass fiber compounds in a subsequent melt compounding and injection molding process, survival of HGMs in masterbatches prepared by in situ reactive extrusion method is substantially higher.

With the advent of high strength HGMs (16,000−30,000 Psi), traditional melt compounding today is a viable technique for compounding HGMs with zero to minimal HGM breakage. Best practices used for other polymers for twin screw melt compounding of HGMs is valid for TPUs as well. These include side feeding HGMs (0.46 g/cc) into a molten TPU polymer melt along with deep cut channels for reduced shear. Table 7.3 compares percent volume loss due to HGM breakage as influenced by HGM loading level, type of feeding (side vs hopper), screw speed, and product form (pellet vs strand) when compounded in a co-rotating twin screw extruder with a standard pelletizer.

Comparing strand and pellet forms, it is apparent that a standard pelletizer causes ample breakage of HGMs in TPUs due to cutting of the soft and stretchy TPU strands in smeared form. Underwater pelletizers prevent HGM breakage in TPUs similar to that seen with other polymers.

Table 7.2 HGM Survival (0.6 g/cc 10,000 Psi) in TPU-HGM Masterbatch Resin Prepared by Traditional Melt Compounding versus Reactive In situ Extrusion Method Followed by Injection Molding

	Traditional Melt Extrusion Compounding + Injection Molding	**In situ Reactive Extrusion Compounding + Injection Molding**	**Traditional Melt Compounding + Injection Molding**	**Traditional Melt Compounding + Injection Molding**	**Traditional Melt Compounding + Injection Molding**
	Masterbatch A	**Masterbatch B**	**Masterbatch A compounded together with glass fiber compound**	**Masterbatch B compounded together with glass fiber compound**	**Masterbatch B compounded together with glass fiber compound**
Glass fiber %			10 wt	19 wt	16 wt
HGM-% 10,000 Psi – 0.6 g/cc	30 wt 45 vol	30 wt 46 vol	11.8 wt 22.5 vol	2 wt	10 wt
Damaged spheres %	17	3–5	37	0	0

Table reconstructed from Ref. [2].

Table 7.3 Volume Loss Percent Due to HGM Breakage in TPUs As a Function of HGM (0.46 g/cc) Loading Level, Type of Feeding, Screw Speed, and Final Form, that is, Pelletized or Strand

	Input TPU Density	Part Density	HGM wt% Loading	HGM vol% Loading	Type of Feeding	Rpm	Form	Input HGM Density	Ash Density	% Breakage
Elastollan 1180A	1.11	1.0588	5	9	Side feed	50	Pellet	0.482	0.530	11.2
Elastollan 1180A	1.11	1.0408	5.2	11.1	Side feed	50	Strand	0.482	0.486	0
Elastollan 1180A	1.11	0.9869	17.7	26.8	Hopper feed	50	Pellet	0.482	0.650	32.0
Elastollan 1180A	1.11	0.9298	17.5	30.9	Hopper feed	50	Strand	0.482	0.526	10
Elastollan 1180A	1.11	0.9233	22	35.0	Side feed	250	Pellet	0.482	0.582	21.2
Elastollan 1180A	1.11	0.8823	22	38.0	Side feed	250	Strand	0.482	0.514	7.9

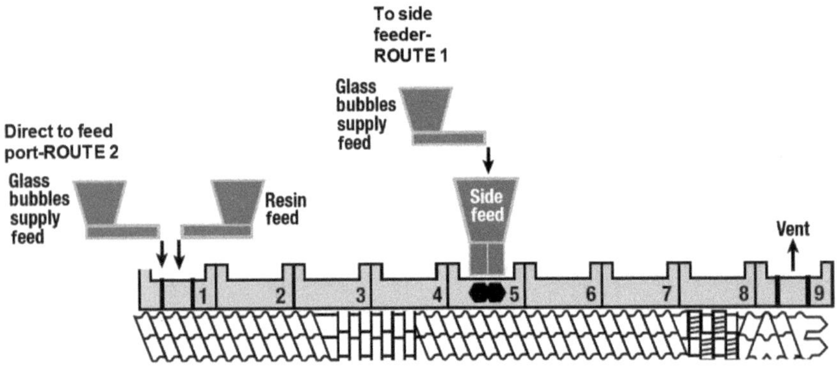

Figure 7.2 Twin screw compounding configuration used to compound TPU/HGMs composites in Table 7.2.

It is also noticeable that side feeding smaller loadings of TPUs and operating at slower screw speeds are more favorable for HGM survival. This is due to reduced shear rates. Figure 7.2 shows the twin screw compounding configuration used to compound TPU/HGMs composites in Table 7.3.

Table 7.4 shows the mechanical properties of thermoplastic linear segmented TPUs with HGMs. Tensile modulus along with stress at 50, 100, 200, and 300% is higher for HGM-filled TPUs. Shore hardness also increases slightly while the density is reduced. Interestingly at 6.5 wt% (13.5 vol%), tensile elongation is still considerable. However, there is decrease in tensile strength at break. Though not shown here, tear strength and abrasion is also negatively influenced by the presence of untreated HGMs. In order to improve tear strength and abrasion, HGM surface treatment in the form of amino and epoxy silane, along with abrasion resistance increasing additives such as isocyanate cross-linkers are useful.

Im et al. surface treated HGMs with aminosilane (HGS-NH2) as well as with TPU grafts (HGS-g-TPU) (Figure 7.3) to develop thermoplastic elastomer composites satisfying the stringent requirements for use as encapsulant materials of underwater devices [3]. TPU/TPU-g-HGM composites exhibited better interfacial adhesion between the TPU matrix and the HGMs than the TPU/HGM composite. As a result, the TPU/HGS-g-TPU composite had better mechanical strength (Figure 7.4) and resistance against swelling (Figure 7.5) with seawater and paraffin oil than the TPU/HGM or TPU/HGS-NH2 composite when the HGM content was fixed. After being impregnated with seawater and paraffin oil, the TPU/HGS-g-TPU composite still maintained better mechanical strength than TPU.

Table 7.4 Mechanical Properties of Thermoplastic Linear Segmented TPUs with HGMs

Component	Formula 1		Formula 2		Formula 3	
	Wt%	Vol%	Wt%	Vol%	Wt%	Vol%
Elastollan 1180A	100	100	93.5	86.5	84.0	72.8
Hollow glass microsphere-iM16K	–	–	6.5	13.5	16.0	27.2
Final	100	100	100	100	100	100
Density	1.11		1.03		0.9619	
Tensile strength at break @ RT (MPa) D412	2800		1720		1020	
Tensile elongation @ RT (%) D412	700		680		335	
Tensile secant modulus@ RT (psi) @ 10% strain D412	2420		3450		5650	
Stress @ 50% (psi) D412	650		800		1090	
Stress @ 100% (psi) D412	770		945		1185	
Stress @ 200% (psi) D412	875		1125		1210	
Stress @ 300% (psi) D412	1015		1300		1095	
Durometer Shore A	82		85		90	

HGMs in Thermoset PU

There are several use areas of HGMs in PU thermosets ranging from reaction injection molded automotive parts to solid syntactic foam insulation coatings to specialty composites and PU foams with improved compressibility.

Reaction Injection Molding of PU

In reaction injection molding (RIM) process, a reactive liquid mixture (usually polyol and isocyanate) is injected or poured into a mold where an

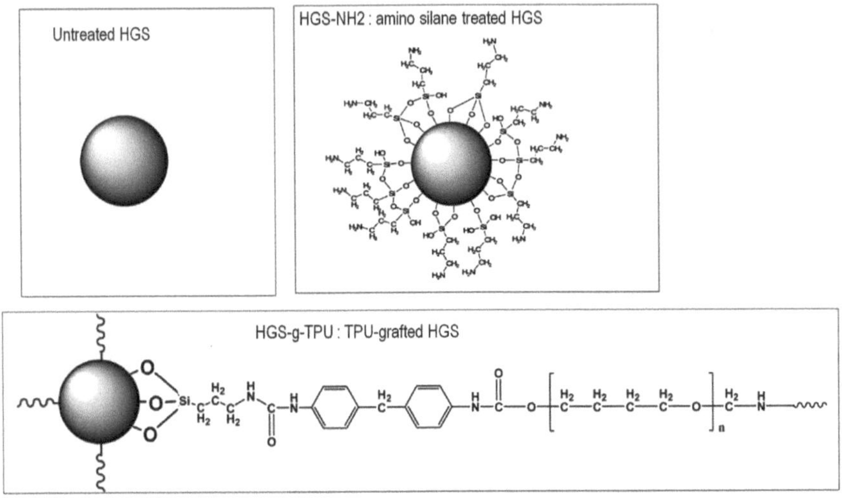

Figure 7.3 Surface treatment options for HGMs in urethane polymers. (HGS is an abbreviation used in the reference and stands for Hollow Glass Sphere. For all practical purposes, HGM and HGS refer to the same material, i.e. hollow glass microsphere.) *With permission from Ref. [3].*

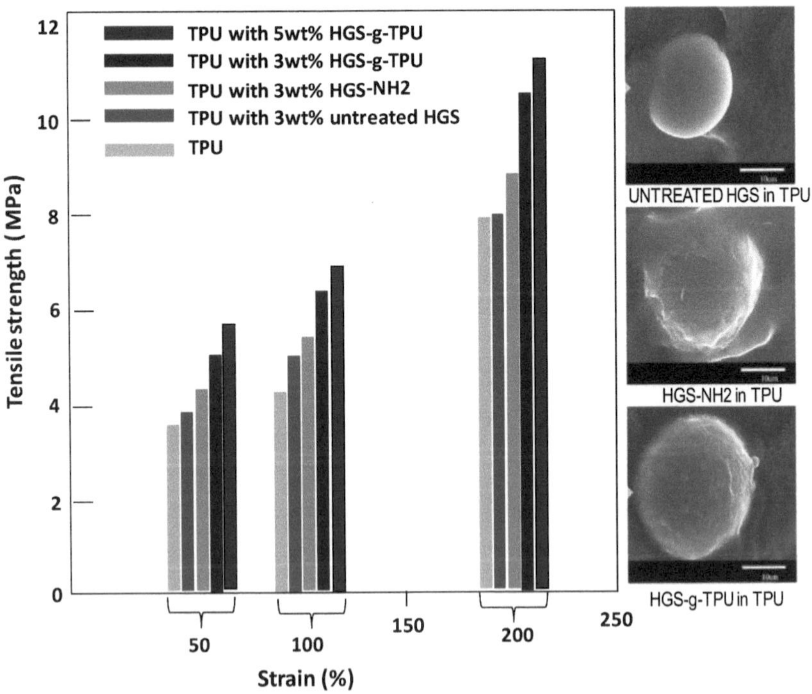

Figure 7.4 Tensile strength of TPU, TPU with 3 wt% untreated HGS, TPU with 3 wt% aminosilane-treated HGS (HGS-NH2), TPU with 3 wt% TPU-grafted HGS, and TPU with 35 wt% TPU-grafted HGS. *Reconstructed from Ref. [3].*

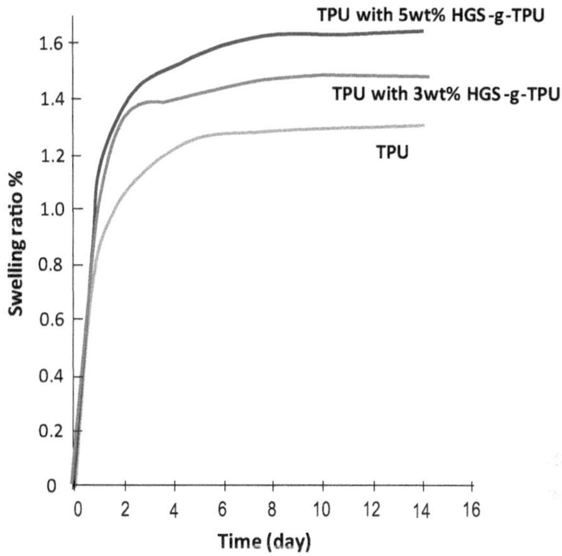

Figure 7.5 Changes in the swelling ratios of TPU composites containing various amounts of TPU-g-HGM as a function of impregnation time in paraffin oil. *With permission from Ref. [3].*

exothermic reaction occurs and the finished part is solidified and removed from the mold. Depending on the chemical formulation, the end product can take on a wide range of physical characteristics: foam or solid, highly rigid or very flexible. Figure 7.6 shows basic RIM process schematics where the two liquid reactants—polyisocyanate component and resin mixture—are held in separate temperature-controlled (32–65 °C) feed tanks equipped with agitators.

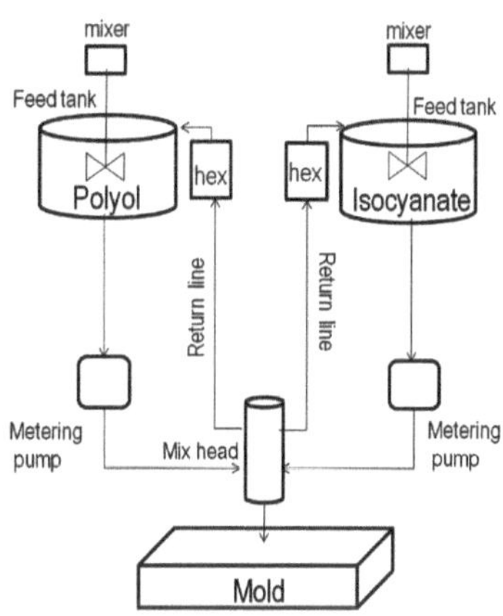

Figure 7.6 RIM process schematics.

From these tanks, the polyol and isocyanate are fed through supply lines to metering units that precisely meter the reactants to the mixhead at high pressure between 1500 and 3000 psi where they are intensively mixed by high-velocity impingement. From the mix chamber, the liquid flows into the mold where it undergoes an exothermic chemical reaction at about atmospheric pressure, forming the final PU article in the mold. While the cycle times are contingent on the complexity and size of the mold and the PU system used, an average mold for an elastomeric part may be filled in 1 s or less and be ready for demolding in 30–60 s after reaching the preset curing time at 66–82 °C. Special extended gel time PU RIM systems allow the processor to fill very large molds using equipment originally designed for molds with smaller volumes.

There are a few important processing points to consider with HGMs in liquid PUs mixing. The pumping system should not exert high shear forces or pressures exceeding the isostatic crush strength rating of the HGMs (see characterization chapter for more details). Positive displacement piston pumps or Moyno pumps are best suited for this purpose. Gear pumps are not recommended because of their high shear and tight tolerances. The metering system should be variable and easily adjusted to accommodate the proper ratio of filled resin as determined by desired HGM loading. The machine should be capable of delivering the desired output at the resin viscosity resulting from HGM loading. Viscosities can be controlled by adjustment of the loading level in each of the components. Pressurized tanks can aid in handling highly viscous components. Heating the tanks and recirculating lines may be advantageous for viscosity control. The mixing head should deliver adequately mixed material without significant HGM breakage from high shear. Rotary mixers run at low to moderate speeds (4000–5000 rpm) are recommended. Static mixers are adequate for formulations where the two components have equivalent and relatively low viscosity. Recirculating systems are required in most applications to maintain uniformity of the filled components. Tank agitators may be required for uniform distribution and suspension of the HGMs in resin.

During the 1960s, RIM was mainly used for manufacturing high-density rigid PU parts such as bumpers and fascias for the automotive industry. Since then, RIM has evolved into an efficient manufacturing process applied to a score of industries: construction, appliances, sports and recreation, electronics, medical, and many others.

Downgauging of RIM bumpers and fascias for the automotive industry was introduced by Dow in 1991 for weight and cost reduction. While gauge reduction reduced the material cost penalty of RIM over TPO at the time, the inherent lower density of the TPOs (~0.92 g/cc) still economically challenged PU materials. The density of nucleated RIM ranges from about 1.02 gm/ccs (unfilled) to about 1.15 gm/cc when molded with 18% fibrous wollastonite filler. In the mid-1990s, James W Berg of Dow Chemical Co reported the ability for RIM fascia to be made both thinner and lighter weight through the combination of reducing wall stock and incorporation of HGMs [4]. This package provided densities of about 0.92 g/cc, equal to those of TPOs. Requirements for painted appearance, durability, and performance were satisfactory. When the use of HGMs was combined with downgauging, up to 40% of the material cost for molding a RIM fascia was typically eliminated.

The mechanical properties of lightweight RIM systems are compared to a typical RIM formulation containing 18% wollastonite in Table 7.5. The low temperature impact and elongation were increased with the lightweight formulations due to the polymer being lighter and somewhat more flexible while the flexural modulus and tear strength were reduced.

Table 7.5 Mechanical Properties for Filled RIM Polymer

Property % Retention Compared to Control	Control 18 wt% Wollastonite	8% RRIMGLOS/ 4 wt% K46 HGM (0.46 g/cc)	12% RRIMGLOS/ 4 wt% K46 HGM (0.46 g/cc)
Flexural modulus	100.0	80.4	79.0
Tensile strength	100.0	96.4	68.6
Elongation	100.0	108.5	87.9
Tear	100.0	76.2	72.6
Heat sag, 6 in	100.0	172.3	151.1
Gardner impact	100.0	102.6	81.1
Gardner impact, −20 C	100.0	420.0	320.0
CLTE, 20–66 C, $\times 10^{-4}$/°C	100.0	100.0	106.0
Water soak, 96 h (% length)	100.0	120.0	92.7
Density	100.0	83.5	82.6

RRIMGLOS: Trademark of Nyco minerals—chemically coupled wollastonite from Nyco Minerals.
K46: 3M Glass Bubbles K46 (0.46 g/cc).
Table reconstructed from Ref. [4].

Heat sag, which is highly dependent on modulus, was increased due to the reduction in modulus. These trends were manageable since the loss of modulus and tear could be offset by adding small amounts of reinforcing filler or by slightly increasing the level of chain extender with no additional cost penalty to the overall formula.

To test the combined effects of lightweighting and downgauging, plaques were molded at a normal gauge of 3.6 mm, an intermediate gauge of 3.2 mm, and at a thin gauge of 2.5 mm. Table 7.6 shows percent property retention of thin wall stocks compared to 3.6 mm normal gauge in HGM containing RIM.

In order to improve the heat sag and water soak, this series was also run with high-performance PU based on a soft segment prepolymer of MDI and significant improvement in heat sag was achieved with less moisture growth. Along with increasing modulus via reinforcing fillers, such high performance polymers were also suggested when downgauging lightweight, microsphere-filled RIM to offset any loss of heat sag resistance.

In this study, the systems were specifically formulated to match the shrinkage of current RIM systems containing either 15 wt% milled glass fiber or 18 wt% wollastonite fiber. From the information in Table 7.7 and Table 7.8, it was possible to match the shrinkage of wollastonite or the newly introduced RRIMGLOS formulations with HGM containing formulation.

Table 7.6 Mechanical Properties for Filled RIM Polymer Conventional Polyurethane +8% RRIMGLOS'''/4% K46 HGM

Property % Retention Compared to Control	3.6 mm Control	3.2 mm	2.5 mm
Flexural modulus (MPa)	100.0	113.8	104.5
Tensile strength (MPa)	100.0	92.6	94.5
Elongation (%)	100.0	88.8	89.9
Tear (kN/m)	100.0	95.3	81.3
Heat sag, 6 in (mm)	100.0	104.9	166.7
Gardner impact, RT(J)	100.0	79.0	50.8
Gardner impact, −20 C (J)	100.0	81.5	76.7
CLTE, 20–66 C, $\times 10^{-4}$/°C	100.0	98.8	96.4
Water soak, 96 h (% length)	100.0	74.2	87.9
Density (g/cc)	100.0	100.0	100.0

Table reconstructed from Ref. [4].

Table 7.7 Plaque Shrinkage As a Function of Wollastonite and HGM

wt% Wollastonite	% Shrinkage Perpendicular 2 wt% K46 HGM	% Shrinkage Perpendicular 4 wt% K46 HGM	% Shrinkage Parallel 2 wt% K46 HGM	% Shrinkage Parallel 4 wt% K46 HGM
5.8	1.1	0.95	0.83	0.7
10	1.03	0.86	0.62	0.52
16.3	0.95	0.76	0.52	0.25

Table reconstructed from Ref. [4].

Table 7.8 Plaque Shrinkage As a Function of RRIMGLOS and HGM

wt% Wollastonite	Perpendicular 2 wt% K46 HGM	Parallel 2 wt% K46 HGM	Parallel 4 wt% K46 HGM
5.2	1.35	0.85	0.68
10	1.27	0.58	0.43
15.2	1.16	0.28	–

Table reconstructed from Ref. [4].

Syntactic PU Foams

Another application of HGMs in PU thermosets is in syntactic foam coating of pipelines used in the recovery of petroleum from the sea. Syntactic foams are composite materials filled with HGMs. Petroleum reserves are increasingly being recovered from great depths. The petroleum from such reservoirs has a temperature of greater than 100 °C (up to 150 °C). This oil is pumped via pipelines from the offshore reservoir to the land. In order to reduce the heat loss from the oil and thereby avoid precipitation of waxes from the oil in the case of a cessation of pumping, the pipeline is insulated with a coating of PU containing HGMs commonly referred to as PU syntactic foam. PU syntactic foam-coated conduit elements are not only classical coated pipes but also objects which are joined to pipelines, for example, muffs, well connections, eruption crosses (also referred to as Xmas tree), pipe collectors, pumps, and buoys.

The process of insulating a pipe with a PU syntactic foam comprises (1) placing a steel pipe in the center of a cylindrical mold (the difference in diameter between the pipe and the mold determines the thickness of the syntactic foam to be applied), (2) pouring a liquid PU reaction mixture containing HGMs around the steel pipe in the mold, (3) allowing the mixture to cure to form a solid PU coating which encapsulates the steel pipe, and (4) demolding the PU-encapsulated steel pipe, thereby yielding a syntactic foam-insulated pipe.

In the preparation of the PU HGM thermoset composites, the HGMs are typically blended with the polyol phase first and reacted with the isocyanate phase, then molded. A continuous method of mixing the hollow microspheres in PU was also demonstrated after the PU system components were reacted with one another [5].

WO 2005/056629 obtained syntactic foam by reacting aromatic isocyanates with a mixture of polyether polyol based on bifunctional and trifunctional primer and a chain extender in the presence of HGMs [6].

A syntactic foam having long pot life (i.e., 4 min) and short demolding times (i.e., less than 10 min), hydrolytic stability, and good physical properties has been reported in EP 1 375 577 A1 [7], US 6,706,776 B2 [8]. These syntactic foams comprised (1) a liquid diphenylmethane diisocyanate, (2) a polyol blend of low unsaturation difunctional polyoxypropylene polyethers, polyoxypropylene polyethers with at least one nitrogen atom and polyoxypropylene polyethers with at least three hydroxyl groups, (3) low molecular weight diols and/or triols, (4) HGMs, and (5) an organometallic catalyst.

The syntactic foams were reported to have high compression E-modulus, greater than 300 psi and could be exposed to water at temperatures of from 0 to 40 °C for up to 10 years without degradation showing good hydrolytic stability. In particular, at about the same hardness and density, a liquid diphenylmethane diisocyanate versus polymethylene poly(phenylisocyanate) in combination with a low unsaturation difunctional polyether polyol, the resultant syntactic foams had higher tensile strength and elongation (about 50% more), higher tear strength (i.e., about 3 times higher), lower Taber abrasion (30–50% less), and considerably lower water absorption (25–50% less in freshwater, and 20–30% less in saltwater).

Specialty PU Composites

PU composite compositions containing 10–20 wt% loadings of HGMs for millable modeling stock applications were reported in U.S. Patent 4,916,173 [9]. These PU syntactic foam compositions had high glass transition temperatures to facilitate their use at elevated temperatures and low coefficients of thermal expansion which were closer in value to that of materials utilized to make composite parts. The materials further showed improved mechanical strength to permit their use under autoclave pressures and temperatures. The millable PU composite was prepared from a polymeric isocyanate, an amine-based polyol, a polyether triol, molecular sieve material (zeolite moisture scavenger), and HGMs. In the compositions of the invention the polymeric isocyanate was present in a concentration of 40–45 wt%, the amine-based polyol 20–25 wt%, the polyether triol 10–15 wt%, the molecular sieve 5–8 wt%, and the HGMs 10–20 wt%. The relationship of the polyols was of particular significance in order to achieve high glass transition temperatures and proper reactivity, that is, control of the reaction.

Lightweight, rapid-setting PU-forming compositions containing hollow microspheres and liquid viscosity reducing agents such as xylene were disclosed in 1973 [10] U.S. Patent 4,038,238. These low-density PUs exhibited the ability to reproduce mold detail in excellent quality. In addition to the excellent mold detail, the products were reported to show less shrinkage in the mold than do the unfilled PUs prepared from rapid-setting PU-forming compositions.

Rigid, flame-retardant PU HGM syntactic foam composite was reported in ref U.S. Patent 4,082,702 [11]. These foams were obtained by mixing an organic polyol, a polyisocyanate, a catalyst, HGMs, and flame-retardant. The flame retardant was substantially colorless with a viscosity of less than 100 cP at 23.9 °C, with a volatility such that it does not evaporate from the reaction exotherm, and with a reactivity such that it was nonreactive in the mixture to the extent that the physical properties of the foam are not substantially changed compared to the physical properties of a foam without the flame retardant. An example of the flame retardant was tris (2-chloroethyl) phosphate. The mixture of the polyol and polyisocyanate was liquid at 25 °C and the amount of HGMs present was sufficient to provide a noncastable mixture in the absence of the flame retardant but castable when sufficient amount of flame retardant was present in the mixture. They also prepared castable mixtures that contained glass fibers to improve the flexural strength. The mixtures prepared cured to a mechanically strong and structural material which was like wood except it was a flame retardant. Table 7.9 shows components of the mixture used to prepare the flame retardant low-density PU syntactic foam composite. This mixture was fluid and could readily be cast into a mold and was nonburning when cured to a rigid PU syntactic foam. On the contrary, when the flame retardant was left out, the mixture was a wet powder, was not castable and when cured, burned. It had a limiting oxygen index (LOI) of 17% oxygen without the flame retardant and LOI of 80% oxygen with the flame retardant. Furthermore, when the mixture contained small amount of glass fibers as shown in Table 7.10, the flexural strength increased up to 35%.

Flexural Strength, kPa

1. 6674 (when 0 part glass fiber was present)
2. 7584 (when 5 part glass fiber was present)
3. 9067 (when 10 part glass fiber was present)

Table 7.9 Formulation of Flame Retardant Polyurethane Compositions

Component	Detailed Description	Parts
Polyol	Organic polyol, formed from the reaction of sucrose, glycerine, and propylene oxide, has an equivalent weight of 160 and a hydroxyl number of 350	100
Isocyanate	Polymethylene polyphenyldiisocyanate	100
Flame retardant	tris(betachloroethyl) phosphate	100
Hollow glass microspheres	HGM	41.7
1:2 triethylene diamine: dipropylene glycol	Extender	1.7

Table 7.10 Fiber-Reinforced Flame Retardant Polyurethane Composition

Component	Detailed Description	Parts
Polyol	Polyether polyol is a 7.0 functional, sucrose/glycerine-initiated polyether polyol	100
Isocyanate	Polymethylene polyphenyldiisocyanate	100
Flame retardant	tris(betachloroethyl)phosphate	75
Hollow glass microspheres	HGM	35
1:2 triethylene diamine: dipropylene glycol	Extender	1.7
Glass fiber	Chopped glass fiber strands of about 6 mm in length	0, 5, and 10

PU Foams

PU foams are prepared by the reaction of liquid isocyanates and polyols in the presence of blowing agents to control density. Water and fluorochemicals are typical examples of blowing agents. In a number of studies on PU foams filled with hollow microspheres, it was shown that the introduction of microspheres could raise the strength and elastic modulus of foams in compression [12–17] and lower the thermal expansion coefficient [13]. It was also reported that the heat and fire resistance of PU foams can be improved [14,15]. However, these effects were considerably dependent on the HGM loading and on the density of the foams made. For instance, it was shown in Ref. [12] that the effect of strengthening of a foam is greater at higher foam densities where the range of foams densities investigated were between 0.160 and 0.450 g/cc. This study is further elaborated below.

Masik et al. filled low-density (0.051 c/cc) PU foams with large particle size (140–160 μm) hollow microspheres of density 0.200 g/cc [15] and reported a gradual increase in the compressive strength and density of the foams case with increasing microsphere content up to 15 wt%.

Yakushin et al. studied the effect of HGMs with a density of 0.125 g/cc on the properties of low-density (0.054–0.090 g/cc) rigid PU foams. In their study, upon introduction of a small (1–2 wt%) amount of HGMs, the characteristics of the PU foam in compression parallel to the rise of direction increased. At a higher content of microspheres, these characteristics worsened owing to an increased amount of defects in the structure of the filled PU foam. The most significant growth (by 10–13%) in the strength and elastic modulus in compression in the foam rise direction occurred in the PU foam with a density of 0.090 g/cc.

However, the elongation at break of the filled PU foam decreased considerably, especially at a foam density of 0.054 g/cc.

In Refs [14,16], it was indicated that the strength of the foams increased with content of microspheres only up to a certain limit. At their greater content, the strength of the filled foams diminished (hollow microspheres of high—0.600–0.700 g/cc—density were examined).

The introduction of HGMs into the initial components of a PU composition changes not only the viscosity of the components but also the kinetic parameters of the foaming reaction. As a result, both the density of the foam and the parameters of its cellular structure can vary [18]. The rigid foam obtained with the incorporation of HGMs differs in appearance from unfilled foam. Unfilled rigid urethane foams have closed cells comprised of gaseous blowing agent enclosed in the PU matrix. Incorporation of HGMs into this material produces a three-phase composite where the volume between the gas filled cells is occupied by a PU matrix containing glass microspheres. Microspheres render smaller foam cell generation.

Table 7.11 shows the improvement in compressive and flexural strength and modulus with the incorporation of HGMs (0.23 g/cc) in PU foams with final densities ranging from 0.160 to 0.450 g/cc. The percentage of improvement in compressive strength when HGMs are added is larger for higher-density foams than low-density foams in this range. For example, it is seen that using foam at a density of 0.168 g/cc with HGMs results in 17% increase in compressive strength whereas the same foam at a density of 0.45 g/cc gives a 48% increase in compressive strength. As for the compressive modulus, approximately 40% improvement is the same for both high- and low-density foams. Flexural strength and modulus show a similar response to density variation as seen in Table 7.12. At a foam density of 0.179 g/cc, HGMs increase flexural strength by 1% while it is 33% increase at a foam density of 0.48 g/cc. Flexural modulus shows an approximately 67% increase with the addition of microspheres for all foams within the density range explored.

HGMs not only directly replace PU resin but also enable lower density foam use at the same performance. For example, flexural modulus of about 50,000 psi can be obtained using 0.3 g/cc foam when HGMs are added, whereas a density of 0.4 g/cc is required for unfilled urethane foam to obtain the same modulus.

As seen in Table 7.13, foam properties can be further improved by varying the urethane composition. If a more flexible urethane foam matrix is formed around the HGM, a further increase in flex mode and strength results.

HGMs also improve foam performance under surface impact. A 1.5-inch thick section of foam at a density of 0.34 g/cc failed at an impact of 90 in-lbs while a similar sample with 9.5 wt% HGMs required an impact of 130 in-lbs to show the similar failure. The low surface alkalinity of the HGMs contributes to the retention of the properties of the filled foams as measured after 1 week at 90 F and 90% relative humidity.

Table 7.11 Compressive Properties of Rigid Urethane Foams with and without Hollow Glass Microspheres Tested Parallel and Perpendicular to the Direction of Foam Rise

	0% Hollow Glass Microspheres					**9.5% Hollow Glass Microspheres**				
Formulation Used	**Density**	**Compressive Strength**		**Compressive Modulus**		**Density**	**Compressive Strength**		**Compressive Modulus**	
		// to Rise	**⊥ to Rise**	**// to Rise**	**⊥ to Rise**		**// to Rise**	**⊥ to Rise**	**// to Rise**	**⊥ to Rise**
	(g/cc)	**psi**	**psi**	**psi**	**psi**	**(g/cc)**	**psi**	**psi**	**psi**	**psi**
Isonate CPR-746-6	0.168	316	373	7880	8006	0.168	367	385	11,100	10,900
Isonate CPR-746-10	0.276	737	702	19,100	18,200	0.285	990	992	29,700	27,600
Isonate CPR-746-16	0.450	2036	1950	44,600	43,000	0.453	2956	3010	63,700	61,000
Stepanfoam HW1260	0.319	830	865	21,600	21,700	0.336	1283	1313	36,000	36,400
Chempol 32-1760/32-1601	0.322	903	860	24,000	21,700	0.315	1140	1137	32,800	31,700

Table 7.12 Flexural Properties of Rigid Urethane Foams with and without Hollow Glass Microspheres Tested Parallel to the Direction of Foam Rise

	0% Hollow Glass Microspheres			9.5% Hollow Glass Microspheres (0.23 g/cc)		
Formulation Used	**Density**	**Flexural Strength**	**Flexural Modulus**	**Density**	**Flexural Strength**	**Flexural Modulus**
		// to Rise	**// to Rise**		**// to Rise**	**// to Rise**
	(g/cc)	**psi**	**psi**	**(g/cc)**	**psi**	**psi**
Isonate CPR-746-6	0.179	440	11,200	0.179	445	18,700
Isonate CPR-746-16	0.479	2160	69,000	0.480	2880	121,000
Stepanfoam HW1260	0.330	1250	39,100	0.330	1410	50,400
Chempol 32-1760/32-1601	0.384	1530	45,050	0.386	2000	76,100

Table 7.13 Variation in Flexural Properties with Polyol Content (Stepanfoam HW1260)

Sample Designation	Amount of Polyol	Amount of Isocyanate	Amount of HGMs	Density	Flexural Strength	Flexural Modulus
	g	g	%	(g/cc)	psi	psi
A	100	100	0	20.6	1250	39,140
B	100	100	9.5	20.6	1410	50,440
C	105	100	9.5	21.9	1510	61,560
D	110	100	9.5	22.5	1570	73,300

References

[1] H. Janik, M. Sienkiewicz, J. Kucinska-Lipka, "Polyurethanes" Handbook of Thermoset Plastics, third ed., Elsevier, 2014, pp. 253–295.

[2] Wilson, et al., Moldable Low Density Thermoplastic Composite with Hollow Glass Microspheres and the Method for Compounding, May 21, 1991. US 5,017,629.

[3] H. Im, S. Chul Roh, C.K. Kim, Fabrication of novel polyurethane elastomer composites containing hollow glass microspheres and their underwater applications, Ind. Eng. Chem. Res. 50 (2011) 7305–7312.

[4] W. James, Berg lightweight reaction injection molded polyurethane for automotive fascia. SAE 9500552 International Congress and Exposition, February 27–March 2, 1995.

[5] P. Huntemann, Method for Producing Syntactic Polyurethane, October 18, 2005. US 6,955,778 B2.

[6] P. Huntemann, Syntactic Polyurethanes and Their Utilization for Off-Shore Insulation, June 23, 2005. WO2005056629 A1.

[7] H. Peter, Markusch, et al., Syntactic Foams with Improved Water Resistance, Long Pot Life and Short Demolding Times, September 9, 2009. EP1375577 B1.

[8] H. Peter, Markusch, et al., For Insulating Pipes, March 16, 2004. US6706776 B2.

[9] Otloski, et al., Polyurethane Syntactic Foam Modeling Stock, April 10, 1990. U.S. Patent 4,916,173.

[10] E. Thomas, Craveus, Low Density Rapid-Setting Polyurethanes, July 26, 1977. U.S. Patent 4,038,238.

[11] R. Jack, Harper, Flame Retardant Rigid Polyurethane Syntactic Foam, April 4, 1978. U.S. Patent 4,082,702.

[12] E. Barber, J. Nelson, W. Beck, Improving properties in rigid urethane foams using hollow glass microspheres, J. Cellul. Plast. 13 (November 1977) 383–387.

[13] J.A. Hagarman, J.P. Cunnion, B.W. Sands, Formulation and physical properties of polyurethane foam incorporating hollow microspheres, J. Cellul. Plast. 21 (January 1985) 406–408.

[14] X.-C. Bian, J.-H. Tang, Z.-M. Li, Flame retardancy of hollow glass microsphere/rigid polyurethane foams in the presence of expandable graphite, J. Appl. Polym. Sci. 109 (3) (2008) 1935–1943.

[15] I.V. Masik, N.V. Sirotinkin, S.V. Yatsenko, S.V. Vakulenko, Effect of glass microspheres on the properties of rigid polyurethane foams, Plast. Massy 1 (2002) 41–46.

[16] P. Yu Blinnikov, E.V. Bykova, A.A. Dorofeyev, G.Kh. Korshunova, Effect of ash microspheres on the properties of rigid polyurethane foams, Tekhnol. Metals 3 (2008) 30.
[17] L.P. Varlamova, V.A. Izvozchikova, A.S. Averchenko, Yu.D. Semchikov, S.A. Ryabov, Effect of aluminosilicate microspheres on the physicomechanical and rheological properties of rigid polyurethane foams, Zh. Prikl. Khim. 81 (3) (2008) 502–504.
[18] O.G. Tarakanov, I.V. Shamov, V.D. Alperin, Filled Plastic Foams (in Russian), Khimiya, Moscow, 1988.

8 Hollow Glass Microspheres in Plastisols

Baris Yalcin

Background Information

Plastisols are relatively stable fluid dispersions of finely divided plastic resin particles in a liquid plasticizer with a small amount of diluent (solvent). Further additives are introduced to the plastisols, such as fillers, pigments, adhesion promoters, rheology auxiliaries (separation inhibitors), heat stabilizers, blowing agents, reactive (capable of cross-linking) additives, and water-absorbing substances (calcium oxide). An organosol differs from a plastisol in that much lower levels of plasticizer and higher levels of diluents (solvents) are used to provide sufficient fluidity to make a fluid dispersion. When lower levels of plasticizers are used, films with much greater hardness can be obtained.

When plastisol solutions are heated above the glass transition temperature of the resin, the resin is solvated by diffusion of the plasticizer and the particles in dispersion fuse (fluxing) into a rubbery plastic mass. Further heating above the melting temperature of the resin creates a homogeneous melt and ultimate strength when cooled to form a solid (curing). The solid is typically tough and elastic especially when the end use is an underbody coating.

The most well-known class of plastisols are poly(vinyl chloride) (PVC) plastisols. PVC is a vinyl polymer that can be produced with different stereoregularity. PVC is usually synthesized by either emulsion (dispersion) polymerization to produce very fine particles (paste grade PVC) or by suspension polymerization to produce larger size particles (dry blending grade PVC). The paste grade PVC from emulsion polymerization is particularly useful for plastisols and is used in higher concentrations than dry blending grade PVC. Vinyl polymers used in PVC plastisols are homo- and/or copolymers of vinyl chloride and other unsaturated compounds, such as vinyl acetate, vinylidene chloride, or vinyl propionate dispersed in a liquid plasticizer or a mixture of plasticizers that are from adipates, sebacates, benzoates, phosphates, phthalates, isophthalates, terephthalates, and polyesters. Typically phthalate type

Hollow Glass Microspheres for Plastics, Elastomers, and Adhesives Compounds
http://dx.doi.org/10.1016/B978-1-4557-7443-2.00008-6

plasticizers are combined with an epoxy such as epoxidized soybean oil which has heat and light stabilizing properties. A particular vinyl resin/plasticizer combination and amounts of each influence the end properties. PVC plastisol coatings are fluxed after application to the substrate, for example, by baking in an oven at 100–175 °C for 10–90 min.

PVC plastisols are used for the most diverse purposes. They are used among other things as underbody coating for motor vehicles, sealing materials for gluing flange seals and for sealing seams from other joining processes, for impregnating and coating substrates of textile materials (e.g., as a coating for carpet backing), as cable insulation, in shell construction in automobile construction, for lining reinforcing structures, such as engine bonnets, boot lids, doors, and roof constructions, etc. Plastisols based on methacrylate copolymers or styrene copolymers also exist and used for such applications.

PVC plastisols as underbody coatings in the automobile industry are widely used because they provide corrosion resistance, road noise suppression, and protection from stones and other road debris. The coatings are relied upon to protect the metal from corrosion by forming a barrier against water and road salt. This feature is especially important as the automobile industry lengthens the terms of corrosion and rust-through warranties.

Example fillers used in plastisols are calcium carbonates, hollow microsphere fillers, talc, clays such as kaolin and china clay, quart, barium sulfate, various fibers, silica, etc. Calcium carbonates fillers are most widely used in vinyl plastisols as resin extender and to increase hardness and tensile strength, impart opacity and whiteness, reduce plasticizer mobility and to control rheology via viscosity build and provide thixotropy. Precipitated calcium carbonate (PCC) with smaller particle size and narrower particle size distribution provides higher increase in viscosity, better thixotropy, whiteness, and reinforcement than ground calcium carbonate (GCC) but combinations of PCC and GCC are also used. Kaolin clay in vinyl plastisols has low abrasivity (hydrous kaolin) and provides improved electrical resistance (calcined kaolin) to plastisols. Barium sulfate in vinyl plastisols increases density, improves chemical resistance, imparts X-ray opaqueness, high load bearing properties, and increases frictional resistance.

Hollow microsphere fillers are mainly used to reduce density while also providing viscosity control, sag, and impart thixotropy to the unfluxed plastisol solution increasing, for example, shelf life of low-density seam sealers (ref). Plastisol-based automobile coatings and seam sealers must meet rigid standards set by various automobile manufacturing companies.

A very important requirement for plastisol coatings is that they be lightweight. As plastisol manufacturers strive for lighter and hence more fuel efficient vehicles, hollow glass microspheres (HGMs) have become a crucial part of the plastisol formulations as key raw material filler. We will discuss HGMs in plastisols in more detail in this chapter.

Typical pigments in PVC plastisols include titanium dioxide (white), iron oxide (red), and carbon black (black). Pigmentation is important not only to achieve the desired color but also to prevent degradation of the vinyl resin from the effect of ultraviolet light.

Adhesion promoters are added to the plastisols in order to affect long-term adhesion of the plastisols on steel, aluminum or galvanized, and/or electrodip-coated or pretreated metal sheets. Basic compounds, such as polyaminoamides (PAA) based on dimerized fatty acids and low molecular weight di- or polyamines, for example, are used as adhesion promoters for PVC plastisols. Basic vinylimidazoles, which are polymerized in as co-monomers, are conventionally used as adhesion promoters for polymethacrylate plastisols. PAA can also be added to PMMA plastisols as an additional adhesion promoter component. In order to render plastisols adhesive to different materials, unsaturated, epoxy, and nitrogen-containing compounds, phenolic resins, and other additives may be used.

Rheology auxiliaries such as separation inhibitors are added to the formulation to improve the stability of the plastisols. Typically separation inhibitors include microcrystalline waxes, hydrogenated castor oil, bentonite, aluminum stearate, fumed silica, and waxes. Plastisol viscosity determines the choice of separation inhibitors used. Separation inhibitors also retard the rise (floatation) of the HGMs to the top in plastisols compositions with low viscosity. This could be a problem if end users do not have facilities for mixing and deaerating at the site of application.

Rheology control agents used as diluents (thinners) in PVC plastisols and organosols include liquid hydrocarbon diluents such as aliphatic hydrocarbons, and mixtures of aliphatic with aromatic hydrocarbons (ref as described in Chapter 26 in volume III of a text entitled Encyclopedia of PVC Technology). Diluents provide lower viscosity and provide better flow and leveling of the plastisol coating.

Plastisols require the use of heat stabilizers to protect the vinyl resin against degradation during the fusion bake. Heat stabilizers are usually combinations of metal salts of organic acids in combination with epoxidized oils or liquid epoxy resins. Calcium-based stabilizers, mixed metal stabilizers (e.g., barium/zinc) are typically used as heat stabilizers in plastisols.

Sometimes, substances affecting the processing behavior are introduced into plastisol formulations. For example, the addition of a powderlike polyethylene (15%) decreases the percolation of a plastisol through stockinet, whereas calcium or magnesium oxide absorbs moisture which worsens the properties of plastisols and the related products. Organosilicon liquids reduce the surface tension of plastisols and thus facilitate a more rapid removal of air bubbles.

Favorable flow properties of plastisols in their unfluxed state at room temperature enable them to be applied to substrates in various ways including dipping, cast molding, rotational molding, extrusion, spraying, spreading, and screen printing. Plastisol gasket materials may be molded, extruded, or calendered. For practical purposes, plastisols can be classified as low viscosity (10–50 P), middle viscosity (50–150 P), and high viscosity (150–10,000 P) at 1 s^{-1}. Different processing techniques require the use of certain plastisols that show certain viscosity behavior as a function of shear rate. For instance, the dipping technique used for manufacture of gloves requires low to middle viscosity plastisols whereas the extrusion technique applied for fabrication of wire covering and elastic profiles may require plastisols with a viscosity of 150–180 P at slow extrusion rates (10–100 s^{-1}) or plastisols with viscosities ranging from 200 to 250 P at higher shear rates (1000–10,000 s^{-1}). Spraying process is employed with plastisols that display reduced viscosities from 10,000 to 110 P with an increase in the shear rate from 0.1 to 150 s^{-1}.

Spray application is one of the widely used methods to apply plastisols especially for automotive underbody coatings. Spray application is typically performed by automated assembly line equipment in place at the automobile manufacturing plants. The sprayable plastisol coating must not excessively abrade the nozzles, pumps, etc. of the spray equipment to prevent costly frequent nozzle replacement.

Figure 8.1 shows a basic outline of plastisol manufacturing steps while Figure 8.2 shows Original Equipment Manufacturer (OEM) typical PVC plastisol layout used in automotive applications. We will discuss below in more detail HGM considerations in the manufacturing steps as well as application steps.

HGM Use and Benefits for Plastisols

As mentioned in previous sections, low density is the major driver for the use of HGMs in plastisols. However, in addition to low density, HGMs also render plastisols with

1. uniform void volume which is necessary for promoting good seal quality
2. thixotropy and sag resistance
3. maintaining a workable low viscosity with prolonged aging without undergoing excessive viscosity increases or undesirable separation of components at room temperature

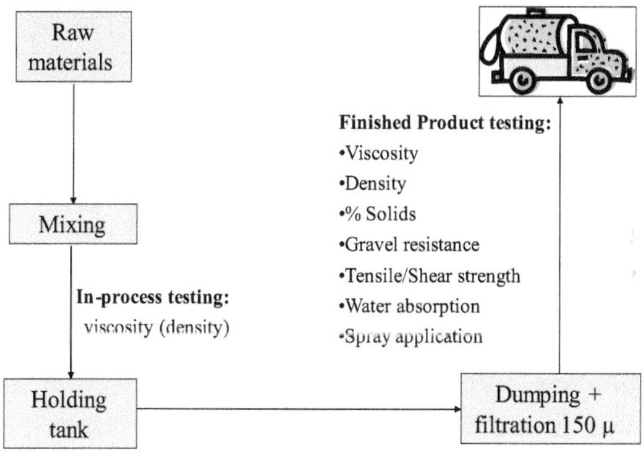

Figure 8.1 Basic outline of plastisol manufacturing steps.

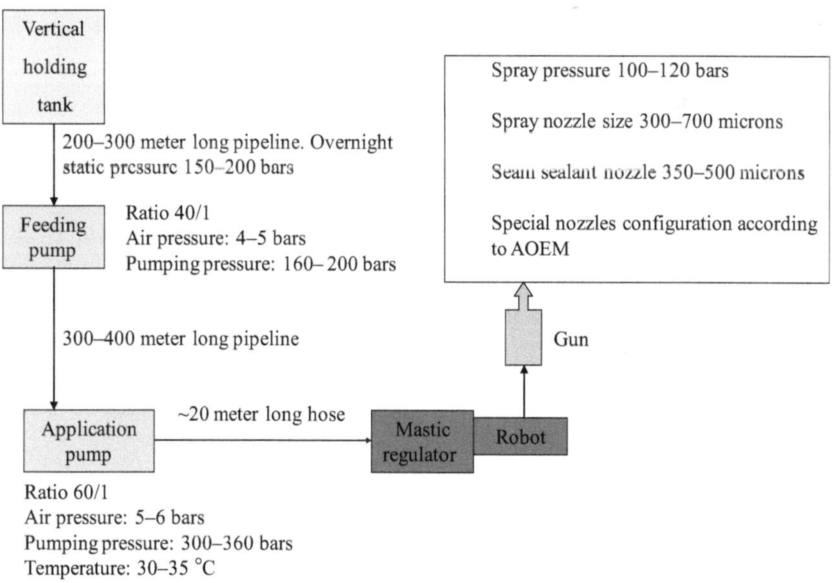

Figure 8.2 Automotive OEM typical poly(vinyl chloride) plastisol layout.

HGMs, especially fused soda lime borosilicate glass, are able to impart plastisols these useful properties because of their low density and high strength which is preserved even while passing through small openings of the spray nozzles without rupturing. HGMs also have excellent chemical and water resistance for plastisol applications.

Table 8.1 shows generic high-density plastisol composition (Formula 1 @ 1.48 g/cc) containing $CaCO_3$ and four other formulations where the density is ultimately reduced to 1.16 g/cc by the addition of HGMs that have a density of 0.38 g/cc. Comparing Formulas 1−5, one can notice that the total filler content ($CaCO_3$ + HGM) is increased from about 25 vol% to about 35 vol%. HGM containing plastisol formulations can tolerate higher filler loadings since HGMs have lower resin demand due to their low surface area and perfect spherical shape. This is shown in Figure 8.3 with volume % of HGM, $CaCO_3$, and total inorganic filler ($CaCO_3$ + HGM) as a function of density.

There are various plastisol compositions containing HGMs in the literature. They were mentioned as early as 1966 in US 3,247,158 [1]. Table 8.2 displays a low-density plastisol composition containing HGMs and a blocked compound of a diisocyanate polymer (i.e., poly-isocyanurate) with styrenated phenol as an adhesion improver as well as a powder curing agent having a melting point of 50−150 °C [2]. Such plastisol compositions are reported to display improved adhesion properties when gelated under low-temperature conditions such as at 110−130 °C for 15−30 min.

A low-density plastisol sealing composition with HGMs which exhibits increased shelf life is shown in Table 8.3 and Table 8.4 [3].

Table 8.3 compares the effect of resin type, that is, dispersion (emulsion method) versus mass polymerized, on the separation of hollow glass spheres (HGS) from plastisol solution as reported in ref, Formula 1 that utilize 100% PVC resin that was prepared by the dispersion (emulsion) method exhibits superior stability without separation of low-density spheres which is essential for good seal quality.

In the same reference, the effect of HGS particle size, that is, diameter, on separation in plastisol was reported (Table 8.4). As the diameter of the HGS decreases and as the density increases, the degree of separation of the spheres from the plastisol also decreases. This relationship was explained by Stokes Law which suggests that the separation rate is directly proportional to the product of (1) the difference in density between the microspheres and the plastisol medium and (2) the square of the radius of the microspheres, and is inversely proportional to the viscosity of the plastisol.

Table 8.1 High- and Low-Density Plastisol Compositions

			Formula 1		Formula 2		Formula 3		Formula 4		Formula 5	
			1.48 g/cc		1.40 g/cc		1.32 g/cc		1.24 g/cc		1.16 g/cc	
	Component	**Density**	**Weight**	**Volume**	**Weight**	**Volume**	**Weight**	**Volume**	**Weight**	**Volume**	**Weight**	**Volume**
		g/cc	**PHR**	**%**	**PHR**	**%**	**PHR**	**%**	**PHR**	**%**	**PHR**	**%**
Resins, etc.	PVC resin	1.39	100	19.34	100	18.76	100	18.18	100	17.60	100	17.02
	Plasticizer	0.99	157	42.77	156	41.49	157	40.21	156	38.92	157	37.64
	Diluent	0.80	33	11.15	33	10.81	33	10.48	33	10.15	33	9.81
	Additives average	1.10	6	1.31	5	1.27	5	1.23	6	1.19	5	1.15
Fillers	$CaCO_3$	2.71	257	25.44	235	22.67	214	19.91	189	17.15	165	14.38
	HGM	0.380	–	–	8	5.00	15	10.00	23	15.00	32	20.00

PVC, poly(vinyl chloride); HGM, hollow glass microsphere; PHR.

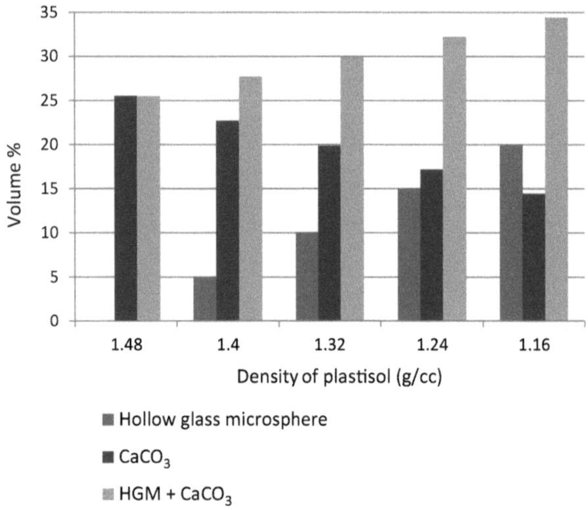

Figure 8.3 Volume % of hollow glass microspheres, $CaCO_3$, and total inorganic filler as a function of density.

Table 8.2 Blocked Polyisocyanurate-Containing Plastisol

Raw material Concentration	PHR-Parts
PVC (Kaneka PCH-12Z)	60
PVC (Zeon G51)	40
DOP	30
Blocked isocyanate compound (TDI blocked with styrenated phenol)	50
Surface-treated curing agent (Nobacure 3721)	2.4
Surface-treated calcium carbonate (Neolite SP)	25
Hollow glass powder (Glassballoon Z-37)	6
Kerosene	15
Calcium oxide	10

PVC, poly(vinyl chloride); DOP, dioctyl phthalate; TDI, toluene diisocyanate; PHR, per hundred resin.
Data from Nakata, Y.; Kunishiga, T., US Patent 4,983,655, January 8, 1991 [2].

Table 8.3 Effect of Resin Type on Separation of Hollow Glass Spheres in Plastisols [3]

Resin, PHR	Formula 1	Formula 2
D642 homopolymer (dispersion)	100	60
GP3-86200, mass polymerized		40
Plasticizer, PHR		
Dioctyl phthalate	68	69
Hollow glass sphere, PHR		
3M C15/25O (low alk)	6	6
Viscosity, Brookfield cps @ 60 rpm		
Initial, room temperature	5720	3650
Separation after 2 weeks		
Room temp	No	Yes

Table 8.4 Effect of Hollow Glass Microsphere Diameter on Separation in Plastisols [3]

Resin, PHR	Formula 1	Formula 2
D642 homopolymer (dispersion)	100	100
Plasticizer, PHR		
Dioctyl phthalate	78	79
Heat stabilizer, PHR		
Calcium-zinc stearate	1	1
Zinc stearate	1.5	1.5
Separation inhibitor, PHR		
Paraffin wax, (127–130 °F melt pt)	3	3
Hollow glass sphere (HGS), PHR		
HGS 1	10	–
HGS 2	–	14.7
Density of HGS g/cc	0.15	0.22
Diameter of HGS microns, approximately	80	40
Viscosity, Brookfield cps @ 60 rpm/110 °F	1590	1710
Separation after prolonged aging	Slight	None

Table 8.5 Effect of Hollow Glass Microsphere on the Void Volume of Fluxed Sealing Material

Resin, PHR	Formula 1	Formula 2
D642 homopolymer (dispersion)	100	100
Plasticizer, PHR		
Dioctyl phthalate	65	65
Heat stabilizer, PHR		
Calcium-zinc stearate	2	2
Separation inhibitor, PHR		
Paraffin wax, (140–145 °F melt pt)	10	10
Chemical blowing agent, PHR azodicarbonamide	3.3	–
Hollow glass sphere (HGS –0.15 g/cc –80 μm), PHR	–	13.6
Viscosity, Brookfield cps @ 60 rpm/110 °F	1590	1710
Effects of flux temperature on the void volume of plastisols	Void volume %	Void volume %
Fluxed @ 300 °F	-1.0 ± 0.6	24.8 ± 0.0
Fluxed @ 325 °F	-1.9 ± 0.8	24.5 ± 0.3
Fluxed @ 350 °F	-0.9 ± 0.3	24.4 ± 0.6
Fluxed @ 375 °F	-2.4 ± 3.2	25.1 ± 0.2
Fluxed @ 400 °F	3.2 ± 3.1	24.9 ± 0.2
Fluxed @ 425 °F	32.0 ± 5.1	25.2 ± 0.2
Fluxed @ 450 °F	50.1 ± 5.4	25.4 ± 0.2

From the results of Table 8.5, it can be seen that the presence of HGS in the plastisol brings about uniformity in the void volume of fluxed sealing material over a wide range of fluxing temperature. When a chemical blowing agent is used, there is a greater variation in the void volume of the plastisol fluxed over a range of temperatures. As noted earlier, uniformity in void volume ultimately results in an improved seal.

Table 8.6 shows a lightweight (1.15 g/cc) and an abrasion-resistant plastisol composition utilizing finely ground rubber useful as sprayable

Table 8.6 Lightweight Abrasion-Resistant Plastisol Composition [4]

Component	PHR
Dispersion grade vinyl chloride/vinyl acetate copolymer (4.9 weight% vinyl acetate content)	100
Plasticizer (mixture of di(C7 to C9-alcohol) phthalates)	241
Ground, previously vulcanized rubber filler (nominal 50 μm average particle size)	50
Long chain hydrocarbon viscosity depressant	30
Wetting agent	7
Adhesion promoter (polyaminoamide)	3
Carbon black	0
Calcium oxide (water scavenger)	16
Zinc oxide stabilizer	1
Calcium stearate-coated calcium carbonate (fine filler, 0.05 μm nominal particle size)	114
Glass microsphere filler (specific gravity 0.46, average size of 70 μm)	12

automobile underbody coating at a viscosity of about 53,000 cps [4]. The ground rubber filler is chosen such that they are vulcanized so that they are not swelled or dissolved by the plasticizer or any other constituent, and they do not chemically react with any constituent because swelling of the filler impairs the viscosity stability and hence shelf life of the plastisol.

Plastisol Mixing and Preparation

Figure 8.1 shows the basic outline of the plastisol manufacturing steps. Mixers are selected according to the materials, viscosity, and batch size. Multishaft mixers are used to mix plastisol solutions with viscosity greater than 30,000 cps. Multishaft mixers address heavy product viscosity and light powder introduction issues via independently controllable shafts with blades that work for a specific mixing function.

A trishaft mixer that combines high, intermediate, and low-speed shaft is shown in Figure 8.4. High-speed shaft utilizes a dispersive blade which

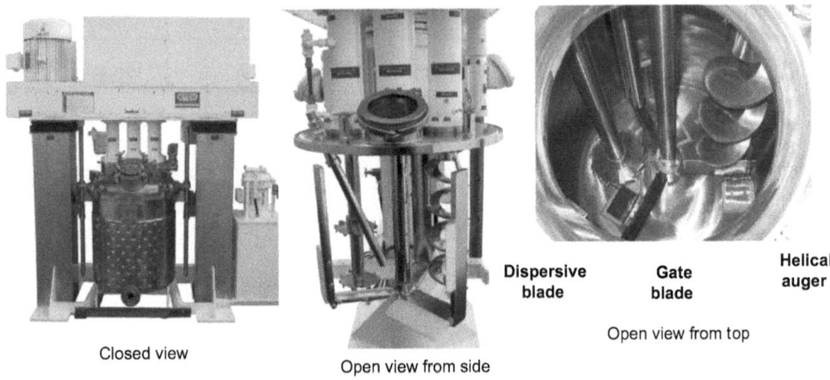

Figure 8.4 Trishaft mixer used for plastisols. *Courtesy of Myers Engineering, Inc.*

is used to break apart agglomerates of fillers and facilitate rapid infusion of the ingredients. Intermediate-speed shaft uses a helical auger to provide gentle mixing and low-speed blade is a heavy duty 3-Prong Sweeping Gate Blade (optionally with helical ribbon) that provides vertical movement at tank wall. The sweeping blade pushes and folds the material into the center as it scrapes product from the inner tank walls, feeding it toward the high-speed shaft. High-speed dispersive blade runs typically in the 3000–4000 fpm range while the intermediate- and low-speed blade runs in the 900–1200 fpm and 400–600 fpm, respectively.

When manufacturing plastisols containing HGMs, sequence of raw material introduction into the mixer in relation to HGMs, mixing time, and speed is important. Sequence of plastisol raw material addition to the mixing machine is important. Typically a portion of the plasticizer is added to the mixing apparatus first along with any compounding ingredients that needs to be dispersed at high speed. Next, the PVC resin is added slowly with agitation while controlling fluidity with additional plasticizer. Generally the remaining plasticizer is added when the plastisol is homogeneous at which point the hollow microspheres are added to the plastisol. The microspheres can be added prior to this point but it is preferred to add them after the plastisol is formed in order to minimize breaking the hollow microspheres during mixing. After the hollow microspheres have been added to the plastisol, the dispersion is mixed until smooth. A mixing time of the order of 10 minutes is generally satisfactory. The resulting mixture is then deaerated and discharged using a ram press. Ram presses, an example of which is shown in Figure 8.5, discharge high-viscosity products such as plastisols that will not gravity feed through the drainage valve.

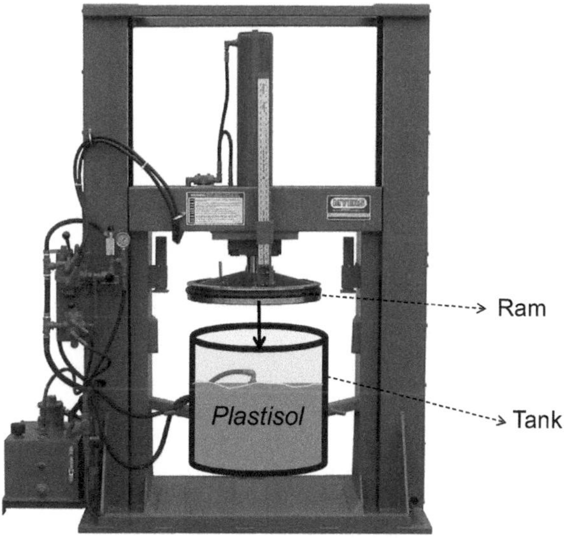

Figure 8.5 Ram Press used to discharge plastisol. *Courtesy of Myers Engineering, Inc.*

References

[1] H.E. Alford, Filled Plastisol Compositions. US 3,247,158, April 19, 1966.

[2] Y. Nakata, T. Kunishiga, Blocked Polyisocyanurate-Containing Plastisol. US Patent 4,983,655, January 8, 1991.

[3] Gibbs et al., Plastisol Sealing Gaskets Puffed with Hollow Discrete Spheres. US 4,485,192. November 27, 1984.

[4] V.M., Deeb, Lightweight Plastisol Coating Compositions. EP0328046 A1, August 16, 1989.

9 Hollow Glass Microspheres in Repair Compounds

Baris Yalcin

Hollow glass microspheres (HGMs) have been a key component of repair compounds for auto, wall, plaster, etc., since the 1960s. In this chapter, we will briefly discuss the use and benefits of HGMs in these application areas.

Auto Repair Compounds

An auto repair compound is used to repair and patch damaged or dented sheet metals. Metal surfaces of automobile bodies (and household appliances) having dents, cracks, or holes are typically repaired with inorganic filler containing thermosetting unsaturated polyester patching compound in order to eliminate these imperfections (Figure 9.1). A satisfactory elimination of such imperfections involves subjecting the dents filled with the hardened patching material to a finishing operation in order to obtain a smooth, uniform surface which blends with the metal surface being repaired so as to be indistinguishable from the rest of the metal surface after a coat of paint has been applied to the surface. Since this subsequent finishing operation is costly and time consuming, minimizing the time required for the finishing operation is of great importance and HGMs play

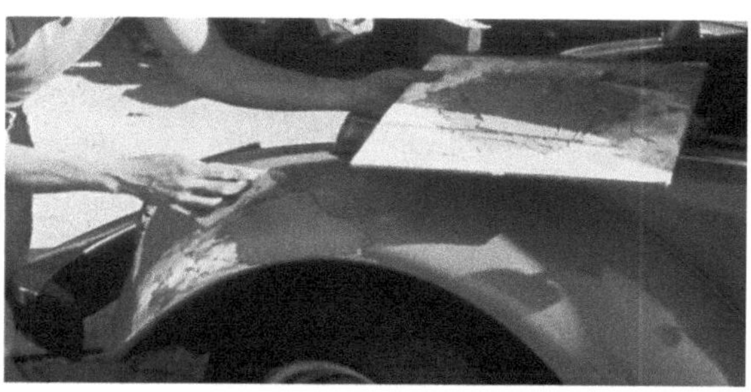

Figure 9.1 Auto repair compound being applied to a sheet metal of automobile bodies.

Hollow Glass Microspheres for Plastics, Elastomers, and Adhesives Compounds
http://dx.doi.org/10.1016/B978-1-4557-7443-2.00009-8

an important role in reducing the time required for finishing operation. In addition, HGMs render the patching material readily workable, that is, ability to file and sand, with good featherability, that is, ability to blend in with adjacent metal surfaces.

Unsaturated polyesters are the typical binder ingredient of an auto repair compound. The unsaturated polyesters used in the patching compounds are usually dissolved in styrene monomer, usually 10–30 part styrene to 90–70 parts by weight of the unsaturated polyester. We have discussed the use of unsaturated polyesters in fiber reinforced sheet molding compound (SMC) applications in a prior chapter. As described in Chapter X, the unsaturated polyester itself is usually formed by the esterification of glycols with dicarboxylic acids or acid anhydrides. One obvious difference in a repair compound versus structural articles made by SMC (or the like) is the absence of glass fibers in a repair compound formula. Typical ingredients of an auto repair compound comprises

- unsaturated polyester resin
- fillers (talc, HGMs, and sometimes calcium carbonate)
- thixotrope and flow agent (e.g., cellulose acetate butyrate (CAB))
- hardener—added just before using

The mixture of thermosetting polyester resin binder/flow additives and inorganic fillers has the consistency of a paste. The filled polyester patching paste and the hardener (cross-linking catalyst such as benzoyl peroxide, cumyl peroxide, and methyl ethyl ketone peroxide) components are packaged in separate containers and blended together (0.5–5 wt%) immediately prior to application on a dented surface to accelerate hardening of the patching compound. At the time of application, the filled thermosetting polyester patching compound is partially gelled and thixotropic and can be formed and shaped to the configuration of the dent, and upon cure, the patching compound becomes a hardened thermoset mass of resin which is confined within the dent.

Table 9.1 shows the ingredients of a metal patching compound (0.77 g/cc and 220,000 cps peak viscosity) where the fillers were incorporated in the polyester resin in major amounts, for example, the total amount of filler particles incorporated in the polyester resin constitutes one-half to two-thirds the total volume of the patching composition [1]. This compound was prepared by first mixing uPS and cellulose acetate butyrate (CAB) in a low speed spiral mixer at 80 °C until a clear solution is obtained and cooled to 20 °C. TiO_2 and the fillers (talc and HGMs) were added subsequently.

Table 9.1 Metal patching compound (0.77 g/cc and 220,000 cps peak viscosity) US Pat. No. 3,873,475 [1]

	Parts
Polyester:styrene solution (70:30)w/accelerator dimethyl aniline	58.7
Cellulose acetate butyrate (CAB)	1.8
TiO_2 rutile	1.0
Talc (10 μm)	23.0
HGM (0.2 g/cc)	15.5

HGM, hollow glass microsphere.

After mixing to produce a uniform blend, a vacuum of 635 mm Hg was applied and mixing continued until a smooth, viscous composition, homogeneous in appearance and free of air bubbles formed. The final compound density was 0.77 g/cc and it had a viscosity of 220,000 cps.

In another early reference [2] (US 4,053,448), it was shown that the proper selection of high density fillers (silica to talc ratio) was a critical feature of the formulation to provide superior workability properties. Table 9.2 shows the ingredients of the auto body repair compound discussed in this reference.

The effect of varying amounts of amorphous silica/talc filler admixture on the performance of auto repair compound is shown in Table 9.3. It shows that the silica/talc ratio of 1:1 to 1:2 gives the best results as far as adhesion, fileability, and sandability. Adhesion was a measure of firm bonding of a sufficiently hardened compound to the metal surface 7 min after catalyst

Table 9.2 Auto repair compound comprising silica/talc high density filler admixture and HGMs (US 4,053,448) [2]

	Parts
Polyester:styrene solution (70:30)	42
TiO2 rutile	1.0
HGM (0.23 g/cc, 500 psi) low density filler	7
Silica/Talc high density filler admixture	50
Benzoyl peroxide	4

HGM, hollow glass microsphere.

Table 9.3 Performance of auto repair compound with varying silica/talc filler admixture

	Filler Admixture		Gel			Elapsed		Elapsed	
	Silica	**Talc**	**Time**	**Adhesion**	**Fileability**	**Time**	**Sandability**	**Time**	**Paintability**
	%	**%**	**Min**	**Rating**	**Rating**	**Min**	**Rating**	**Min**	**Rating**
1	33	17	2.5	Good	Good	7	Good	12	Excellent
2	25	25	2.5	Excellent	Excellent	7	Excellent	12	Excellent
3	17	33	4.5	Excellent	Excellent	10	Excellent	14	Good
4	8	42	5.0	Good	Good	12	Good	20	Poor
5	42	8	2.5	Poor	Poor	7	Poor	12	Excellent
6	50	0	2.5	Poor	Poor	14	Poor	20	Excellent
8	0	50	5.5	Poor	Poor	14	Poor	20	Poor
9	46	4	2.5	Poor	Poor	7	Poor	12	Excellent
10	4	46	5.5	Poor	Poor	14	Poor	20	Poor

Table 9.4 Ingredients of an auto patching compound useful for repairing galvanized steel

Ingredients	Wt%
uPS resin	33
Styrene monomer	15
Toluene sulfonamide–formaldehyde resin	3
Talc	44
HGM	5

HGM, hollow glass microsphere.

addition. Fileability was a measure of the ease of hand filing the hardened patching composition, that is, if the hardened patching composition was soft and pliable enough to be hand filed, after 7–14 min. Sandability was a measure of whether or not the patching composition applied to the dent can be sanded to trade specifications after 12–20 min have elapsed from the time of catalyst addition to the patching composition. Paintability was rated based on appearance being free of pinholes and blisters.

In reference US 4,980,414 [3], it was shown that a uPS-based patching compound that includes aryl sulfonamide-aldehyde resin could be used for repairing metal surfaces with galvanized steel (Table 9.4). When the metal requiring repair is galvanized steel, it is difficult to ensure adhesion of the body filler to the galvanized steel surfaces. Generally, the galvanized steel surfaces are pretreated in order to provide the desirable adhesion of the body filler to the galvanized steel. The pretreatment generally involves thoroughly grinding the galvanized surface to remove the zinc coating, sanding of the surface, acid etching of the surface to remove any remaining galvanized coating, neutralization of the acid, followed by cleaning and drying of the surface. In many instances, this treated surface is coated with an epoxy/urethane primer to ensure corrosion protection (since the galvanized surface has been removed) followed by light sanding of the primer. The body filler can then be applied to the pretreated galvanized steel surfaces, and the adhesion of the body filler to the surface is generally acceptable. However, this process, which is required to prepare galvanized steel for body fillers is both time consuming and costly. Also, the corrosion protection is lost by removal of the galvanized surface provided by the primer which is often not the equivalent of the corrosion protection afforded by the original galvanized surface.

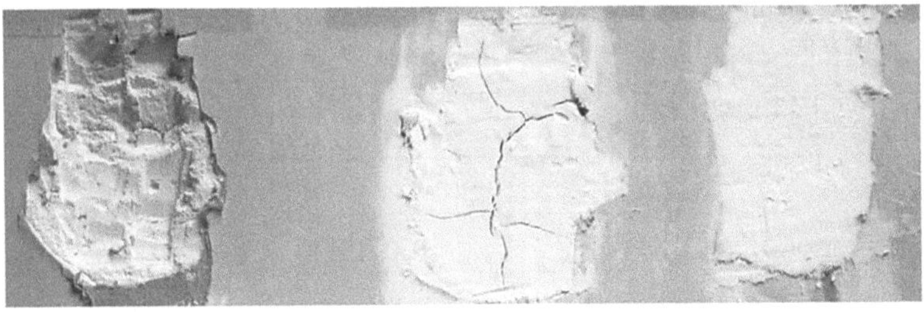

Large hole in plasterboard

Heavyweight spackle

Lightweight spackle

Figure 9.2 Heavyweight versus lightweight spackle in plasterboard.

Wall Repair (Spackle Compounds)

Interior walls of buildings are often constructed using gypsum wallboard panels (sometimes referred to as drywall, plasterboard). Where cavities, recesses, holes, etc., may be present (due to imperfections or damage) it is common to use wall repair compound (often referred to as "spackling") to fill such cavities (Figure 9.2). Conventional wall repair compounds often include

1. polymeric resin binder(s) (e.g., polyvinyl alcohol (PVA) emulsion in water)
2. thickeners
3. water
4. inorganic filler(s)
5. other additives (biocide, antifreeze)

Binders

Binders are often supplied as an aqueous latex emulsion (40–60% solids of polymeric resin binder, in water). Polyvinyl acetate (PVAc) polymers and copolymers are the well-known polymeric resins suitable for binders for wall repair compounds. Other resins such as ethylene vinyl acetate polymers, styrene-butadiene polymers, polyacrylamide polymers, natural and synthetic starch, natural rubber latex, and casein can also be used as binders alone or in combination with other binders. Ideally, the polymeric resin binder should have a glass transition temperature (Tg) of around room temperature (from about 20 °C to about 30 °C). A Tg in this temperature range renders the binder well suited for fusing and coalescing under ambient conditions after the wall repair compound has been applied and allowed to dry. If the Tg is excessively higher than RT, the binder might be below its Tg in the dried compound and render the compound relatively brittle and prone to cracking. Conversely, too low a glass transition might render the wall repair compound too soft or rubbery for sanding. It is also beneficial if the binder has a relatively broad Tg range (~ 10 °C) such that the binder does not exhibit a relatively sharp change in physical properties upon changes in ambient temperature.

Thickeners and Cothickeners

Conventional wall repair compounds often comprise organic polymeric thickeners. Such organic polymeric thickeners are often used to provide an increased viscosity of the wall repair compound so that the compound does not excessively sag, slump, or run when applied to a vertical wall. They are often designed to exhibit their thickening effect by their interaction with the water that is present in the wall repair compound. Therefore, they are often water soluble or water swellable (around 22 °C) due to the presence of multiple hydroxyl groups in their structure. They are commonly referred to as water retention agents in the industry. The most well-known polyhydroxy compounds are cellulose ethers (e.g., methyl cellulose, hydroxypropyl cellulose, hydroxypropyl methyl cellulose, hydroxyethyl cellulose, hydroxyethyl methyl cellulose, hydroxyethyl hydroxypropyl cellulose, ethylhydroxyethyl cellulose, and sodium carboxymethyl cellulose). Such thickeners can also include, for example, polyethylene glycol, polyethylene oxide (and/or polyethylene oxide/ polypropylene oxide copolymers), PVA polymers or copolymers.

Synthetic Inorganic Fillers

Certain natural or synthetic inorganic fillers (e.g., clays such as attapulgite, bentonite, montmorillonite, illite, kaolinite, sepiolite), while not necessarily water soluble, are known to exhibit a thickening (e.g., viscosity-increasing) effect when dispersed in water. Such materials (particularly those that absorb water and/or swell upon exposure to water) have commonly been used as thickeners in wall repair compounds (they are also occasionally referred to as rheology modifiers, nonleveling agents, slip agents, etc.), and are known to contribute to shrinkage upon drying as discussed in US 4,824,879 [4]. These clay fillers also work as a cothickener with cellulosic, associative, and alkali swellable thickeners.

Water

Wall repair compounds often comprise a significant amount of water (e.g., greater than about 20% by weight), such that, after the wall repair compound is applied to a wall, the water evaporates over a period of time resulting in the formation of a dried, hardened material which can be sanded, painted, etc.

Inorganic Filler(s)

Inorganic fillers used in wall repair compounds include talc, $CaCO_3$, and low density fillers such as HGMs. Because of the use of HGMs, the spackling are lightweight and provide less trowel drag with great non-leveling characteristics. One other advantage to using HGMs is that there is minimal shrinkage. At 66% by volume, the spheres are packed in as tightly as possible. Therefore, even when the water is removed, the system as a whole does not move. Talc cannot be packed as tightly so when the water is removed, there is movement between the talc particles resulting in shrinkage. Another benefit is reduced drying time. With the lower water level, the drying time is reduced and the amount of labor is reduced.

The use of HGMs in a spackling composition was shown earlier in references US 3,386,223 [5] in 1968 and US 4,391,647 [6] in 1983. A spackle composition containing HGMs from US 4,391,647 [6] is shown in Table 9.5. The composition was reported to be also suitable for use as a stucco material, for example, in refurbishing ceilings and it provided fire resistance and insulating values.

Borate-containing HGMs were reported to cause quick gelling in the presence of PVA binder reacting with the hydroxyl groups [7]. Although the problem was more pronounced when PVA was used as a binder alone, it was also present to a lesser degree when PVA was added in certain amount in the PVAc emulsion. Guar gum, a naturally occurring high molecular weight polyhydroxy compound, similarly gels in the presence of borate. Gelation may result in other cases where high molecular weight polyhydroxy compounds are present in the wall repair compound used for water retention, viscosity control, latex stabilization, etc. Although gelation may be counteracted by acidifying the system, other problems could result. For example, fillers such as calcium carbonate can dissolve in acidic systems, releasing carbon dioxide and thus causing intolerable bubbling. US 4,629,751 [7] selectively employed a low molecular weight polyhydroxy compound in a borate-containing environment to prevent an undesirable reaction with a high molecular weight polyhydroxy compound. The composition shown in Table 9.6 was reported to be free of

Table 9.5 Spackle composition containing HGMs US 4,391,647 [6]

	%	%	%
Water	47.9700	47.2	50.0
Tamol SN surfactant	0.8722	0.9	0.99
Triton X-100 surfactant	0.4361	0.5	0.5
Attapulgus clay (cothickener)	3.4887	5.0	
Methocel (water retention agent)	0.7414	0.8	0.8
Dowicil 75 (anti microbial) 1-(3-chloroallyl)-3,5,7-triaza-1-azoniaadamantane chloride	0.0044	0.01	0.01
Talc	7.8496	8.0	8.8
Mica	2.4421	2.6	2.7
Marble white	27.9098	29.99	31.0
Polyvinyl acetate	4.3609	0.8	0.8
HGMs (0.15 g/cc −D50 50 μm)	3.9248	4.2	4.4
Reported benefits	Excellent adhesion to the spackling knife, little or no trowel drag, good workability, lightweight, nonleveling properties, and good mixability with normal drying time and tape adhesion.	Exhibited excellent workability, slip, resistance to flow, good tape bonding, and minimum cracking	Same as second but with better adhesion to holes and crevices

HGM, hollow glass microsphere.

Table 9.6 Spackle composition utilizing low molecular weight polyhydroxy compound (sorbitol) US 4,629,751 (1986) [7]

	Parts by Weight	Calculated Vol %, Solid Basis
5% (weight) aqueous solution of PVA (binder)	800	3.68
Ethylene glycol	15	1.38
Preservative	5.5	0.51
Attapulgus clay	18	0.64
Ground calcium carbonate	244	8.33
Talc	77	2.53
Mica powder	36	1.45
Thixotrope (hydroxyethyl cellulose)	5	0.46
Sorbitol (low molecular weight polyhydroxy compound)	12	1.11
HGM (0.15 g/cc, 250 psi)	130	79.91

HGM, hollow glass microsphere.

gelation, even after standing for several days. When applied to gypsum board, the composition was reported to adhere well to seam tape, showing little or no signs of sagging or leveling when applied to vertical surfaces.

Recently, the use of smoothing agents, such as glycol ethers, was taught in US 8,653,158 when preparing a wall repair compound with low shrinkage [8]. Such smoothing agents appear to function by reducing the apparent viscosity of the wall repair compound rather than increasing it, while not causing unacceptable sagging or slumping. The presence of such a smoothing agent was also reported to improve the ability of the compound to be brought back to the above-described smooth consistency by the addition of a small amount of water, in the event that the compound is inadvertently allowed to lose water (e.g., by the container being left open for a period of time). In the absence of such a smoothing agent, the addition of water only served to reduce the viscosity of the compound such that unacceptable sagging or slumping resulted. In the reference, the smoothing agents employed in the wall repair compound comprised a bimodal mixture of substantially spherical synthetic inorganic fillers such as HGMs and ceramic microspheres.

Table 9.7 Low shrink wall repair compound employing low molecular weight glycol ether smoothing agent Lightweight Wall Repair Compounds. US 8,653,158 (2014) [8]

Component	Wt%
UCAR binder emulsion 626	57.32
K20-3M HGM	25.95
Ceramic microspheres	15.94
Propylene glycol Butyl ether	0.24
Polyphase P20T	0.42
Mergal 192	0.11

HGM, hollow glass microsphere.

Glycol ethers are low molecular weight (e.g., from about 90 g/mole to about 250 g/mole) and are typically liquid at room temperature (e.g., 220 C). While being partially or completely miscible with water, they do not substantially increase the viscosity when added to the water phase Table 9.7.

Tape Joint Compound

During installation, gypsum board surfaces frequently develop cracks, pits, etc. When gypsum board is mounted, there are inevitably dimples at the location where individual sheets are nailed or screwed to studs or ceiling joists. A strip of perforated tape is commonly applied over the indented adjacent edges of a space between adjoining gypsum board panels, spreadable joint filling compound being applied both under and over the tape and allowed to dry or cure. Similar to spackling compounds, tape joint compounds include polymeric binder, filler, and water, but also typically include preservatives, water retention agents, wetting agents, defoamers, plasticizers, nonleveling agents, etc.

US 4,824,879 [4] describes a low-shrinkage tape joint compound. Tape joint compounds typically include enough water to permit them to be readily and smoothly applied with a spatula or trowel. A substantial amount of shrinkage typically results during drying of the water. It is thus generally necessary to apply such compounds in several separate thin coats, sanding at least after the last coat has dried, in order to avoid leaving a disfiguring depression where the joint has been filled. For example, when filling taped joints between abutted gypsum board panels, it is usually necessary to apply three coats of a joint filling compound having a shrinkage of 30–40%, drying and sanding between applications. If tape joint compound shrinkage is 20–30%, it may require only two applications.

Tabulated in Table 9.8 is a tape joint compound with low shrinkage, 14%. The low shrinkage was achieved by minimizing the volume percentage of water adsorbing additives such as the conventionally included attapulgite and hydroxyethyl cellulose, thereby obtaining compounds having lower water content. This decreased shrinkage was reported to significantly reduce the need for additional coatings, even where the area to be filled is deep or extensive, and thereby greatly lowers labor cost. In preparing these formulations the attapulgite and "Tamol" dispersants were vigorously mixed with approximately half of the water, using an air motor and a high shear mixer so as to thoroughly disperse the clay. This dispersion was combined with the PVAc (supplied as an aqueous emulsion) and preservative, using a Ross double planetary mixer.

Table 9.8 Low-shrinkage tape joint compound

Component	Wt%	Vol%
Polyvinyl acetate (40% aq emulsion)	1.9	3.0
HGM	0.51	6.0
Attapulgite	0.68	0.5
Hydroxyethyl cellulose	0.12	0.3
Calcium carbonate	70.0	45.4
Nuosept 95 preservative	0.01	0.0003
Tamol 850 dispersant	0.34	0.5
Talc		
Mica	1.6	1.0
Water	24.8	43.3
Density, g/cc	1.8	
Viscosity, Brookfield RVT, spindle F	2800	
Shrinkage %	14	

HGM, hollow glass microsphere.

The remaining dry ingredients were blended by hand and added to the planetary mixer. The ingredients were then mixed for 5 min, adding just enough water to provide a stiff but fluid mass. After the sides of the mixing vessel had been scraped down, the remaining additional water was added to obtain the desired viscosity, on the order of 2000–3000 Pa.s.

Final Word

Repair compounds, whether it is auto, wall, tape, ceiling, or stucco, benefit greatly from the use of hollow glass microspheres due to their low density, excellent spherical packing leading to high solids content with no sagging and low shrinkage, great nonleveling characteristics, and less trowel drag.

References

[1] Pechacek et al., Composition for Filling, Patching and the Like. US 3,873,475, March 25, 1975.
[2] A.W. Holle, Polyester Based Patching Composition. US 4,053,448, October 11, 1977.
[3] P.E. Naton, Plastic Body Filler. US 4,980,414, December 25, 1990.
[4] Montgomery et al., Low Shrinkage Tape Joint Composition Containing Attapulgite. US 4,824,879, April 25 1989.
[5] A.A. Wegwerth, Method of Joining Drywall Panels. US 3,386,223, June 4, 1968.
[6] Deer et al., Spackeling Composition. US 4,391,647, July 5, 1983.
[7] R.L. Montgomery, Gel-resistant Bubble-filled Wall Repair Compound. US 4,629,751, December 16, 1986.
[8] Gozum et al., Lightweight Wall Repair Compounds. US 8,653,158, Feburary 18, 2014.

10 Handling of Hollow Glass Microspheres

Baris Yalcin

Handling of powders and bulk solids has been reported in numerous references in the literature. Characterization of powders and bulk solids regarding their flow properties plays an important role, for example, for product development and optimization and customer support. The discharge of powders and bulk solids from silos, hoppers, transport containers, etc., may result in severe problems due to flow obstructions, segregation, shocks and vibrations, or unsteady flow. To avoid such complications, solutions have to be found considering the flow properties of the bulk solid.

Hollow glass microspheres (HGMs) share several of the common handling techniques and equipments applicable to other bulk solids and powders. However, due to their low density, fragile, and free-flowing nature, some differences exist between HGMs and other bulk solids such as talc and $CaCO_3$. It is the purpose of this chapter to review some of the equipment that has been successfully utilized in the past to transport and store HGMs. It is important to note that HGMs, depending on the manufacturer, may come with different free-flow behavior. The techniques and equipment, that is, hoppers, metering and mixing equipment, and pumps, mentioned here are based on experience with 3M HGMs and were not verified for other HGM brands by the authors of this book.

Silos and Hoppers

In high-volume processing operations, day and weigh hopper combinations are recommended (Figure 10.1). While the day hopper stores enough HGMs to feed one or more weigh hoppers throughout a work period, weigh hoppers are placed close to the mixers to help minimize any delay in getting HGMs into the process. For a smaller processing operation, a single hopper is fed directly from the shipping container functioning both as a day and weigh hopper.

Symmetrical hopper designs can create funnel flow and discharge problems as shown in Figure 10.2. HGMs in symmetrical hoppers could

Hollow Glass Microspheres for Plastics, Elastomers, and Adhesives Compounds
http://dx.doi.org/10.1016/B978-1-4557-7443-2.00010-4

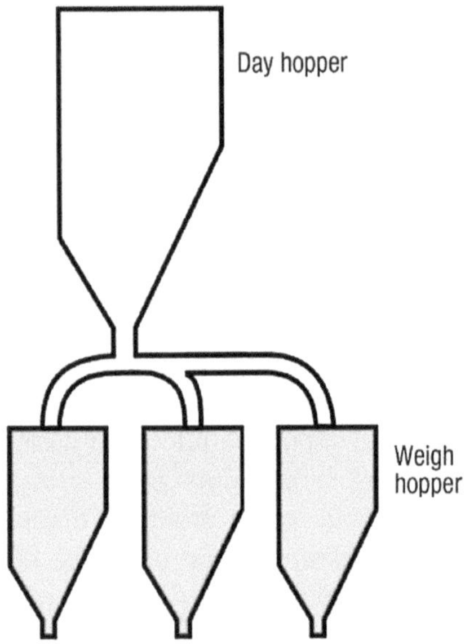

Figure 10.1 Day and weigh hopper combinations [1]. *With permission from 3M.*

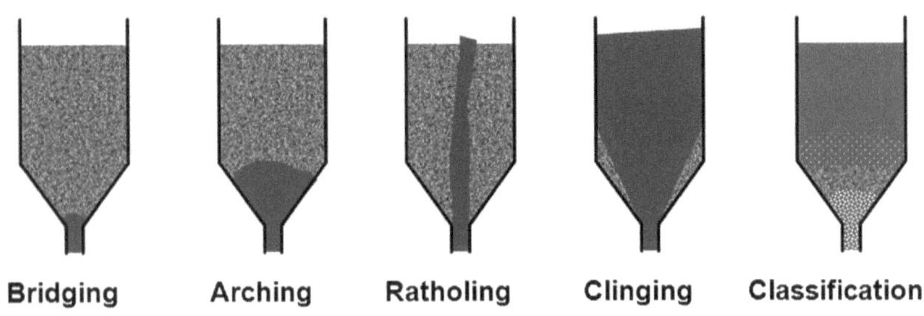

Figure 10.2 Flow issues with HGMs. *Courtesy of 3M.*

bridge or arch, and then rat-hole when discharging. Flow could be erratic and density of flow can vary. Product can remain in dead zones until complete clean out of the system.

To help avoid these problems, asymmetrical day or weigh hopper designs with steep sides and offset entrance, outlet or sloped regions, and flow inserts are recommended (Figure 10.3). Asymmetric hopper design helps improve flow speed and uniformity, while reducing the amount of dust generated with high-volume conveying equipment.

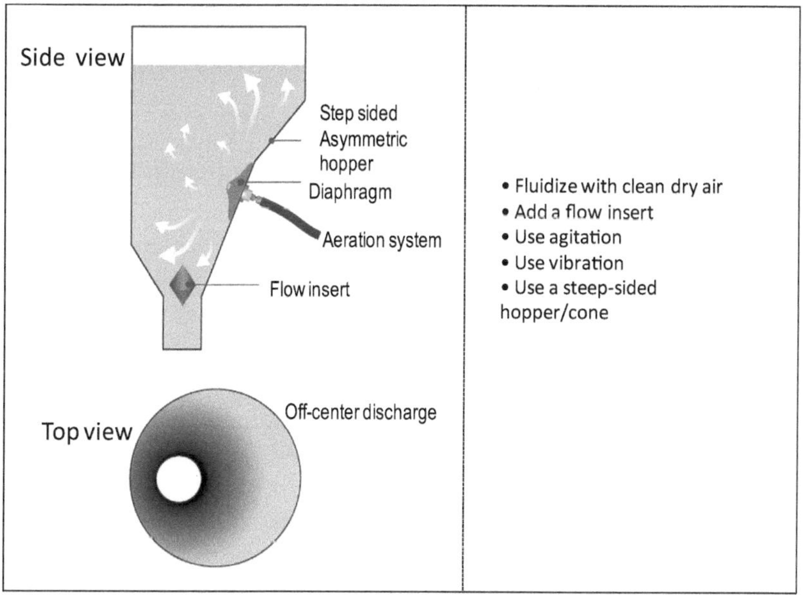

Figure 10.3 Asymmetric hopper design for HGMs. *Courtesy of 3M.*

If an existing symmetrical hopper is to be used, an aeration system to fluidize the HGMs can be incorporated to help mitigate these discharge problems. Fluidizing with clean dry air helps reduce discharge time and HGM breakage. Aeration also helps blend the materials, improving output consistency. Aeration increases bubble bulk volume considerably, so the hopper must be large enough to accommodate the increased volume. A combination of aeration with asymmetrical hopper design provides the most optimal discharge of HGMs from hoppers.

Transfer of HGMs

Typically, boxes are emptied by vacuum suction. Vacuum can be obtained from a vacuum receiver or a double diaphragm pneumatic pump.

Double Diaphragm Pneumatic Pump Transfer of HGMs from a Box or a Bag

A 3-in pneumatic double diaphragm pump is typically used to move lightweight powders. It is a lower initial cost method that effectively transfers aerateable low-bulk density powders. The air-driven pump is a combination of pull/push, vacuum-pressure conveying system. The pump pulls material by vacuum into its inlet, and then pushes the material along the conveying line with pressure. In the pressure conveying system, poor line connections will leak dust into the workplace. The pump should be placed closer to the process in order to pull material a longer distance. This will reduce line plugging. Adding purge air into the pump chamber when the pump is pushing material into the line helps to decrease pump plugging and stalling. The pump should not be stopped when it is full of powder. Often a vacuum relief valve is mounted close to the pump suction port. A bleed-down valve at the pump outlet is suggested for relieving pressure from a plugged line or pump. Purging the pump and the conveying system with air or other compatible gas is suggested before and after HGM transfer. The pump should be operated between 25 and 50 psi and below the hose maximum burst pressure rating (Figure 10.4).

Figure 10.5 shows a double diaphragm pump vacuuming HGMs from a box to an asymmetric hopper and mixer.

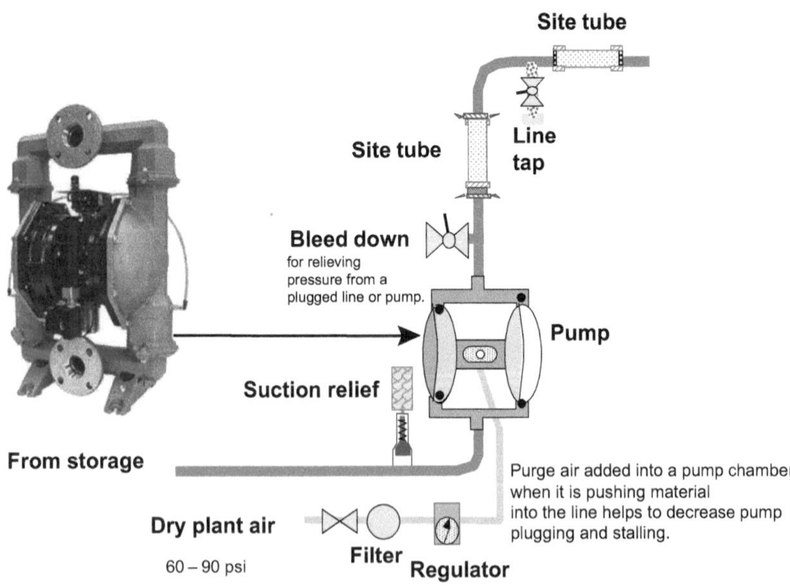

Figure 10.4 Double diaphragm pump system [2]. *Reconstructed from Ref. [2] with permission from 3M.*

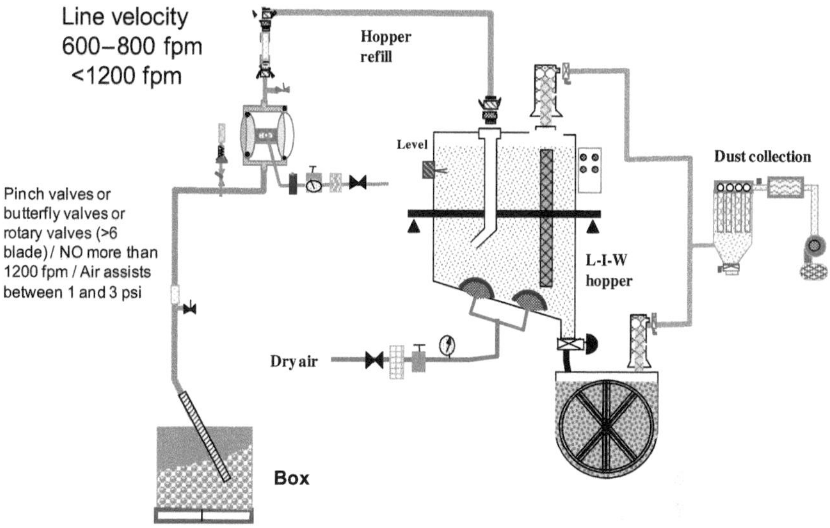

Figure 10.5 Transfer of HGMs from a box using a double diaphragm pump to an asymmetric hopper followed by the mixer (L-I-W = loss-in-weight) [2]. *Reconstructed from Ref. [2] with permission from 3M.*

Vacuum Wand Pickup

Aerated pickup wands are often constructed from a rigid pipe. The wand is fitted with a low-pressure compressed air line to fluidize material near the tip. The wand tip should have a protective guard to reduce plugging with the liners in the box. This enables the system to resume transfer even if the tip is covered with material. The diameter is the same as the hose connection. Fluidizing air must be synchronized with the transfer system operation.

There are several commercially available wands or they can be made easily. Figure 10.6 shows a simple handmade vacuum wand, effectively used with a double diaphragm pump. Such a wand features adjustable air registers for regulating air in outer supply tube and inner fluidizing air channel. As the diaphragm pump creates vacuum to pull the HGMs, the ambient room air is sucked in through the registers of the wand into the inner channel, which aerates the HGMs in the box near the wand tip. The wand should be inserted into an HGM box at one of the top corners and slowly guided diagonally through the microspheres to the opposite bottom corner to create a trench. This will allow material to fluidize as it falls from the trench walls into a pool for pickup by the wand.

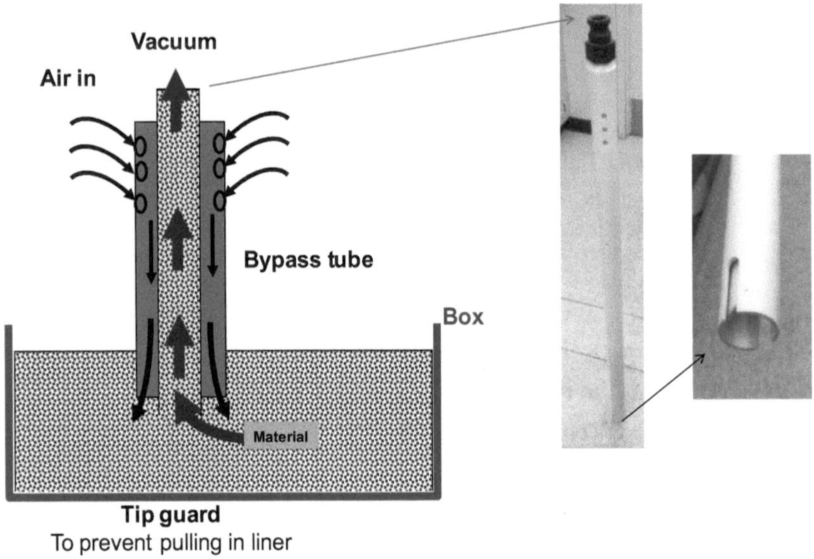

Figure 10.6 Vacuum wand. *With permission from 3M.*

Conveying Lines and Hoses

Conveying lines connect the various system components for microsphere handling. Typically, a transfer system uses 3-in (76 mm) components. HGMs should be transferred with a line velocity of less than 1200 ft/min (365 m/min). Lines with long radius bends or sweeps are suggested instead of 90° elbows. Lines can be combinations of rigid and flexible materials. All conveying lines and all components should be electrically grounded. Hoses with a smooth inner bore and a conductive drain wire are suggested. The drain wire must be connected to metal connectors. Flexible lines may range from braided chemical hose, semitransparent PVC, clear polyurethane to interlocking metal hose. Caution has to be exercised as some of the hose materials are limited to use in temperatures above 20 °F.

Flow Aids

Experience shows that fluidizing a material makes it much easier to handle. Air assists (Figure 10.7) in the conveying lines are used to keep conveying lines trouble-free. They are typically mounted at the bottom of vertical line legs and about every 50 ft in horizontal line runs. Air pads mounted near the discharge port in hoppers are suggested to help fluidize material for easy transfer.

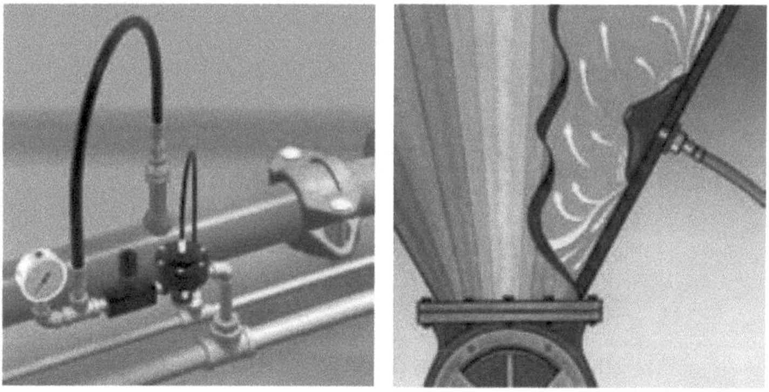

Figure 10.7 Air assist (left) and air pads (right). *Reconstructed from Ref. [2] with permission from 3M.*

Sight Windows and Sight Tubes

Sight windows and sight tubes are used to observe material flow in order to locate a problem in the transfer system. Suggested mounting locations are at the pump outlet or the bottom of vertical legs, or optionally at the receiving vessel entrance. Sight tubes use Pyrex® glass or transparent PVC schedule 80 tubing. Grounding with a wire across the length of the sight tube is suggested. Polycarbonate material is suggested for sight windows.

Tilt Table for Boxes

Large boxes are often placed with the pallet on a tilt table. The table is tilted after a portion of the material has been removed. It may include options for a wand holder and vibration. Excessive vibration will decrease glass microsphere transfer rates.

Vacuum Transfer of HGMs from a Box or a Bag

Another effective way to transport HGMs is by a vacuum transport system (Figure 10.8). As opposed to diaphragm pump transfer, vacuum transfer is a pull-only conveying system that operates at a negative pressure. It may use a venturi, two-stage fan, or a positive displacement blower (illustrated) to move the air that carries the material. The vacuum system will move material at higher line velocities than a pump system. The advantage is that it does not emit particles into the work area. The vacuum system is also not prone to line plugging problems. The primary filter is usually cleaned with pulsed, high pressure, clean, dry air. A secondary

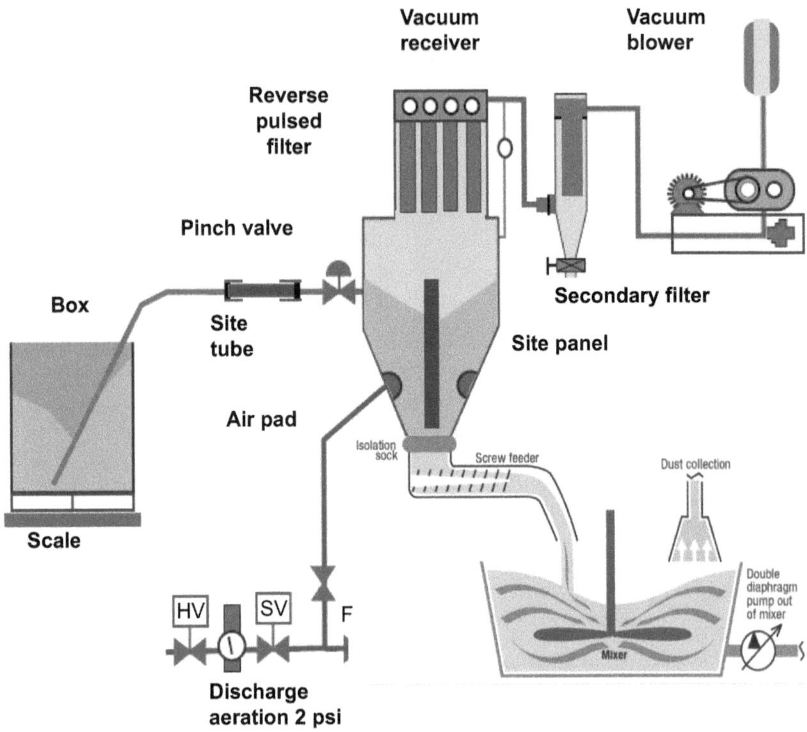

Figure 10.8 Vacuum transfer system. *Reconstructed from Ref. [2] with permission from 3M.*

filter is placed after the receiver filter in order to protect the fan or blower. An adjustable vacuum relief valve regulates vacuum in the receiver. Typical suction is 50−100 in of water column. A sight panel and aeration of the hopper is suggested. Aeration pads are typically offset away from the center axis, near a corner of a pyramidal hopper and at different heights in the transition section. The transition section is the converging section from the hopper vertical section to the discharge valve. With hopper aeration, bubbles will flow easily. This has the advantage of reducing height for fitting into a height-limited area above a mixer. Typically, bubbles are metered continuously to a process using a loss-in-weight screw feeder. Batches processed are charged by weight from a hopper on load cells or by placing the box on a floor scale. It is important that the screw feeder utilizes a variable flight depth and pitch screw as shown in Figure 10.9.

Powders and other bulk solids tend to segregate when the particles are different, in size, shape, and/or density. Most segregating materials are

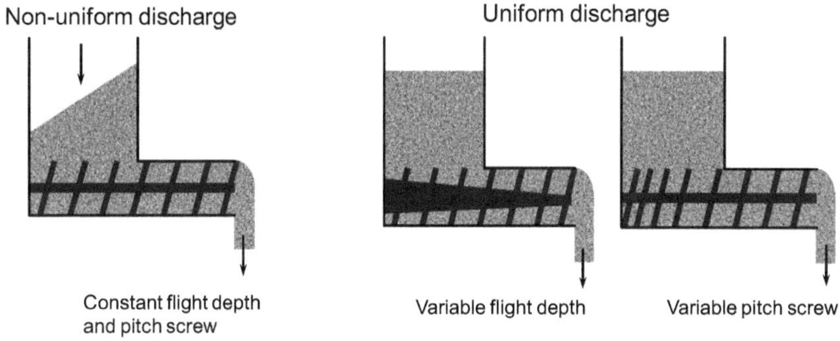

Figure 10.9 Nonuniform and uniform discharge using screw feeder. *With permission from 3M.*

free flowing or slightly cohesive so that the particles can easily separate from each other. On the contrary, the behavior of poorly flowing bulk solids (materials containing fine particles or moisture) is dominated by interparticle adhesion forces, which reduce the mobility of individual particles and, thus, the tendency to segregate.

Usually one wishes to avoid, or at least reduce segregation, since many downstream processes require a product with constant composition. For example, if the bulk solid must be filled into packages for customers in small quantities of equal composition, or if in a process (e.g., grinding or combustion) steady-state conditions can be attained only with a feed of constant composition. Furthermore, segregation can result in inaccurate volumetric dosing due to a fluctuating bulk density.

During use of either a vacuum transfer system or pump, handling of the dusting at the receiving vessel (separation of air and HGMs) needs to be considered. The system would need to be tightly sealed with a dust collector at the receiving vessel using a dust sock or reverse jet pulse pleated filter system.

Bulk Bag Discharging

Bulk bags are a common way of packaging HGMs. The most dust-free method of discharging bulk bags is to place the bag into a bag unloading station. The stations include a glove box hopper with a window and a light and flexible dust containment cover (Figure 10.10). The flexible hopper top is slightly stretched by the bag weight in order to minimize dust when the bag discharge spout is opened. The station can be enclosed on three sides. Usually, dust collection vents are placed above the top of the hopper. If desired, the station can include a weigh cell system for metering

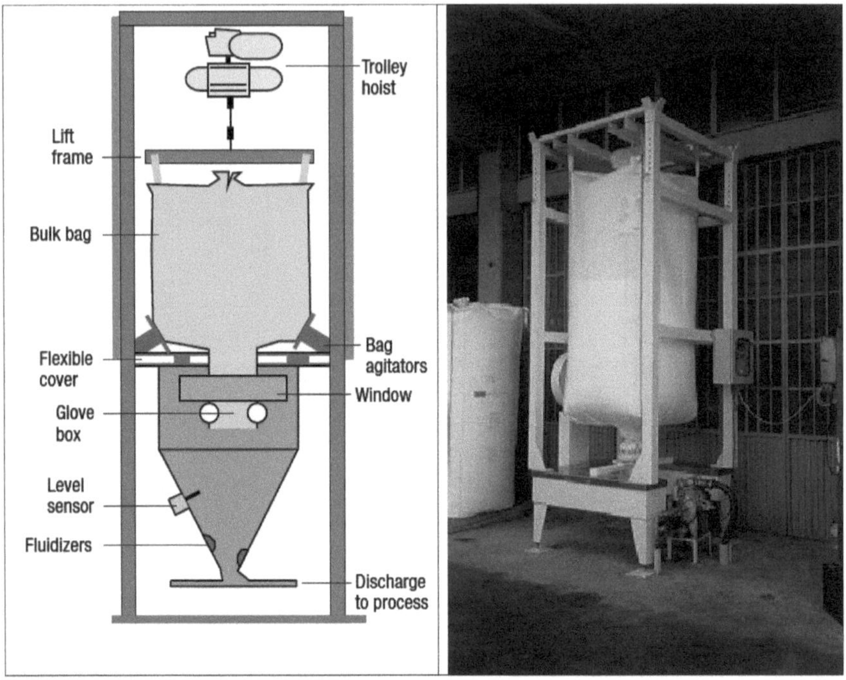

Figure 10.10 Balancing fluidization for pump transfer and feeding into mixing equipment [3]. *Reconstructed with permission from 3M.*

material to the process by weight. Capacitance sensors are used to detect material level. Often for mix tanks, the bulk bag is discharged directly into the entry way of the vessel. This requires an auxiliary filter port in order to collect the dust created during bag discharging. Material is usually transferred by vacuum suction from a vacuum receiver or a double diaphragm pneumatic pump.

In the preceding sections, we emphasized the importance of fluidizing the HGMs via air assists, wands, or air pads for efficient transfer via a diaphragm pump or a vacuum transfer system. Although fluidizing HGMs is crucial for pump or vacuum transfer operations, highly fluidized HGMs may cause problems during their incorporation into mixing systems such as side feeding into a twin screw extruder. Highly fluidized HGMs, when intersected with a viscous melt stream, could be pushed back due to their increased rotational motion in their fluidized state. On the other hand, HGMs with balanced flow have enough translational motion that could overcome the force exerted by the flowing melt stream (Figure 10.11).

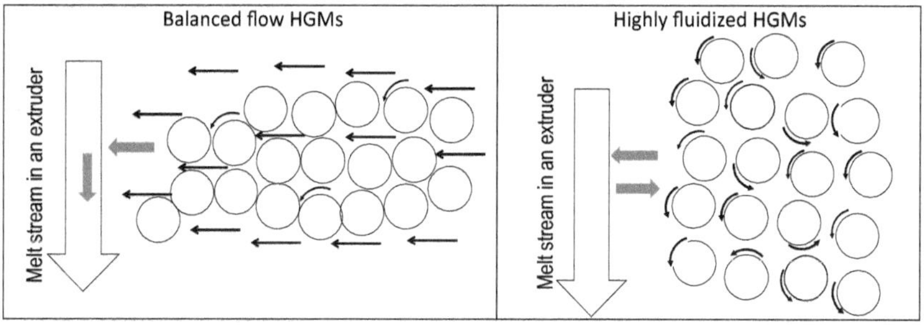

Figure 10.11 Balanced versus fluidized flow of HGMs in a process.

Cohesiveness and fluidizability of HGMs may vary depending on the manufacturer of the HGM. However, for very fluidizable HGMs with long air retention time, a more balanced flow can be achieved by allowing the HGMs to settle for a period of time in day hoppers in order to increase cohesion before being side stuffed. In addition, HGMs free falling from tall silos or hoppers into a side stuffer should be avoided, since free-falling HGMs will readily be fluidized and may cause problems in feeding into melt streams such as in an extruder. Angle of repose for HGMs, that is, angle which is formed by a pile of solid with the horizontal, is about 40°–50° in their packed state but 0° in their fluidized state.

References

[1] 3M Technical Paper. 3M Glass Bubbles Metering and Mixing Guide.
[2] 3M Technical Paper. 3M Glass Bubbles Box Unloading Solutions.
[3] 3M Technical Paper. 3M Glass Bubbles Bag Unloading Solutions.

11 Mixing and Dispersion of Hollow Glass Microsphere Products

Stephanie Shira and Cary Buller

Hollow glass microspheres have helped to revolutionize the area of polymer fillers with an extensive variety of glass microsphere offerings. A diversity of diameters and isostatic collapse strength varieties are available for a range of mixing, processing, and end-use solutions. The advantages offered by hollow glass microspheres are many, yet due to their unique aspect ratio, density, and chemical properties, they must be handled differently than typical mineral fillers, particularly in storage, formulating, metering, mixing, pumping, and weighing. Due to their ultra-low-density, hollow glass microspheres occupy up to 20 times the space for a given weight of a traditional filler. Also, due to their thin-shelled hollow nature, they may be broken by compression at constrain points, shear under areas of tight clearance, and impact at turns in transfer equipment. As a result of the above, one must beware to minimize the compression, impact, and shear forces upon the hollow glass microspheres to ensure they are not damaged during processing and transport.

Hollow Glass Microsphere Transport to Mixer

The first step involved in processing with hollow glass microspheres is getting them to the resin or polymer matrix. Dispensing hollow glass microspheres can be challenging, as using traditional dispensing machines can result in a significant glass microsphere breakage rate. Factors which impact the breakage rate when transporting hollow glass microspheres include material agitation methods, material viscosity, choice of pump, and temperature while incorporating hollow glass microspheres.

One must be cautious of the electrostatic forces encountered when transporting hollow glass microspheres; piping should be grounded to ensure static control. Once grounded, the main concern of transporting and mixing hollow glass microspheres is breakage. To minimize breakage a number of steps are suggested: introduce hollow glass microspheres in

Hollow Glass Microspheres for Plastics, Elastomers, and Adhesives Compounds
http://dx.doi.org/10.1016/B978-1-4557-7443-2.00011-6

a pulselike flow pattern into the vortex below, as well as introduce hollow glass microspheres as late as possible in the mixing process to minimize the chance of breakage. To these ends, a high flow–low-shear mixer should be selected, and it includes anchor, bar turbine, horizontal hydrofoil, pitched blade, plow, propeller, and multishaft mixers. What these types of mixing devices have in common are low dispersive and high distributive mixing characteristics. The key is to avoid use of high shear, as this can cause damage to hollow glass microspheres. Due to the small size and low-density nature of hollow glass microspheres, one should use a dust collection system when transporting hollow glass microspheres into the mixer, as well as in other areas of potential contamination or glass microsphere escape.

When transporting hollow glass microspheres, wet or dry (i.e., before or after mixing), one should use one of the following types of pump: double diaphragm pump, non-intermeshing gear pump, or peristaltic pump. These pumps are ideal because they generally allow for high glass microsphere survival rate at the pump's maximum hydrostatic pressure. One must be sure to select a glass microsphere which has great enough strength to survive the pumps hydrostatic pressure; centrifugal, gear, lobe, progressive cavity, or close-contact intermeshing-type pumps should be avoided, as they may result in significant glass microsphere damage.

Other methods of transporting hollow glass microspheres into the product involve the use of seals, such as a diaphragm pump, or a hose with a stinger. These schemes provide an effective way of moving hollow glass microspheres to the mixer while reducing hollow glass microspheres in the atmosphere around the mixer. A proven and preferred method is to transport hollow glass microspheres into a silo above the mixer and use a rotary valve to control the rate of addition of hollow glass microspheres, such that the mixer is not overfed.

When processing, it is important to avoid pressures greater than that for which the selected hollow glass microspheres are rated. One should avoid high-shear processing with equipment such as high speed dissolvers, roll mills, and ball mills, or processes which may generate point contact shear, such as gear pumps or three-roll mills.

Fundamentals of Dispersion

To have the best understanding possible for handling and mixing hollow glass microspheres, one must first grasp the fundamentals of dispersion. A complication which arises from simply vigorously mixing materials is that the very fine particles (sometimes microscopic in size) still maintain electrical and molecular attraction, and/or mechanical bonding. These fine particles tend to lump together and form agglomerates which no amount of mixing will break. An aggregate (or agglomerate) is composed of a group of particles that are strongly adherent and can be broken down only by the application of relatively strong mechanical forces. In the days before dispersers, stone mills were used to physically grind dry materials into a fine powder. Subsequently, ball or pebble mills were used; one can envision these types of mixers as a type of spray paint can, where the marble within the paint can is used to homogenize the product. The balls or pebbles fall through the product, impacting the agglomerates and breaking up lumps, and thereby coating (or "wetting") each basic crystal so that it does not stick to other particles. This process results in a smooth mixture. A further advancement was made by the introduction of roller mills, which grind the agglomerates until they are reduced to their basic particle size or smaller, and wet them to ensure they do not recombine. The ball, pebble, and roller mills break up agglomerates as well as grind pigments and other particles in the product into smaller basic sizes or crystals.

The above processing techniques were an advancement in their day, but they did take a very long time to produce a smooth product, sometimes in excess of 24 h. With the advent of the disperser, the deagglomeration process could be accomplished much more rapidly, and resulting in a smoother, more uniform end product.

The basic part of the disperser is the impeller, which is a thin disc with carefully designed teeth distributed radially about the circumference. The impeller is much more than a mixer; its hydraulic action tears particles apart and disperses them uniformly throughout the product. This work is done with two actions: firstly, particles hitting the impeller are broken apart, or deagglomerated; secondly, the intense turbulence surrounding the impeller causes particles to hit each other with great momentum and inertia. The energy of this impact physically breaks apart agglomerates. This area of immense turbulence is at the tip of the blade and extending out several inches, and is known as the zone of attrition (the location and size of this zone varies with impeller design and size). Beyond the turbulent zone, the product is processed and thoroughly mixed by the laminar flow created by the impeller. The flow separates upon impact with the tank walls, ensuring complete circulation of the batch thereby effectively homogenizing constituents. While the above processes are imperative to guarantee a homogeneous product, these forces can be damaging to fragile hollow glass microspheres, and should therefore be completed before the addition of constituents which do not require breakup, or which may be shear-sensitive.

It is sometimes necessary to heat the product within the tank to facilitate a chemical reaction; this can be accomplished by increasing the rate at which the impeller rotates, or by utilizing a jacketed tank.

85–90% of dispersers are used to deagglomerate formerly ground pigments or solids, and subsequently disperse these solids throughout products such as chemical compounds, inks, paint, paper coatings, plastics, and rubber compounds. These pigments or particles, as noted above, were previously ground into fine particles; however—especially when dry—the particles tend to come together in a lump and stay as such (i.e., agglomeration) until they are pulled apart again (deagglomeration). This is due to an array of attractive forces, including Van der Waals forces, adhesion, interfacial tension, etc. The intense shearing and turbulence generated by the disperser tears apart these lumps and disperses them into the coating, ink, paint, or plastic at an exceptionally rapid rate. In many cases, a product made with bulky older equipment could take upwards of 24 h; the same product could be made better in 40 min with a modern, compact disperser such as a dual-shaft quad-blade disperser with twin motors (see Figure 11.1). This type of dual-shaft mixer imparts 30% more energy into the batch due to the blades rotating in opposite directions at the point which they overlap, creating opposing flows which baffle each other. This allows for the blades to run faster while avoiding cavitating—the formation of an air pocket which could damage the product, as well as the dispersion equipment.

A second use for dispersers is to grind off fine particles from a solid mass, such as a resin. Lacquers—made from solid resin combined with a solvent—are produced in this fashion, and can be made quickly with a disperser. Approximately 5% of dispersers are used to grind and peel away solid material which is then dispersed.

The remaining roughly 5% of dispersers are used for food cutting, mixing foodstuffs into a smooth, homogenized paste. An impeller with serrated edges and sharp angular teeth is used to create the necessary turbulence, shearing, and mixing motions—which are similar to those of a standard impeller used for manufacturing applications. This type of cutter blade is sometimes used to break up chunks of resin in the making of lacquer, as previously mentioned.

There has been a trend in the mixing and dispersing industry toward greater uniformity and consistency of products. A dispersion equipment manufacturer has developed systems that monitor, report, and track historical information to greatly aid in documenting and certifying

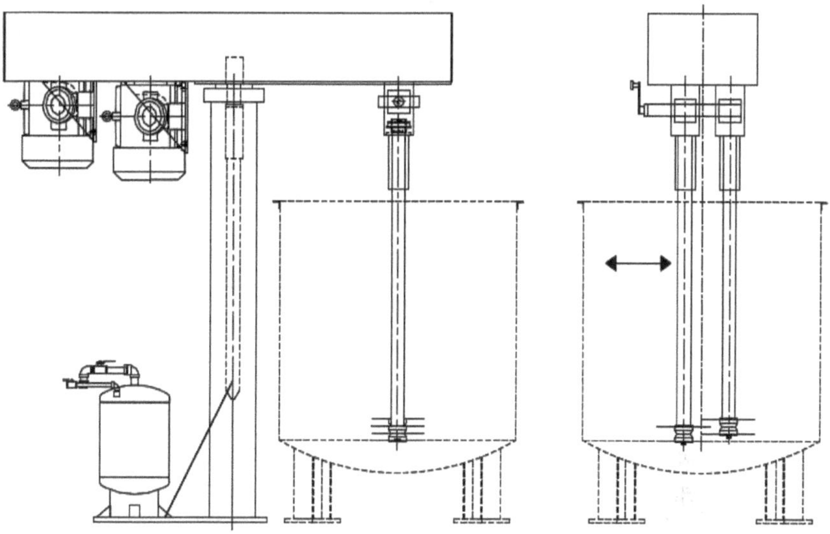

Figure 11.1 A basic design for a dual-shaft disperser. Two motors allow each shaft to rotate at independent speeds. Blades can be "open" or "closed" (see front view) to change intensity of shear, size of vortex, and flow/mixing patterns. *Courtesy of Myers Engineering, Inc. (Myers 850 Series).*

batches. Removal of the human element—and thus variability from one operator to the next—from the production process results in vastly improved product consistency from batch to batch, or from one production facility to another. By custom designing a mixer, one can better ensure repeatable results. The location and rate of additions can be more exacting with a custom machine, and is incredibly important for interbatch homogeneity.

As a case study, a mixer and disperser manufacturing company's customer, a manufacturer of mastics at five sites throughout the United States, had a corporate-level issue with batch production. The plant in Texas was requiring 45 additional minutes to produce a batch, as compared to a plant in Southern California which produced the batch in the desired time frame. The disperser company's technical team was called upon to determine the reason for this delayed mixing. The plant in Southern California was adding their calcium carbonate through a silo immediately upstream of the high speed vortex. The calcium carbonate then did not sit on the batch very long; it rotated only 90° before being pulled into the high-shear zone. It was discovered that the manufacturing plant in Texas was adding calcium carbonate to the opposite side, downstream of the vortex, 270° from being pulled into the vortex. This forced the calcium carbonate to travel three-quarters of the way around the tank before entering the vortex, allowing the calcium carbonate to sink and settle, and leading to agglomeration. The agglomeration was the cause of the excess mixing time; to break up the agglomerates which formed in a short span of time—less than 2 min—required a significantly lengthened mix time. This demonstrates the importance of location of addition, and its effect on mix time, as well as homogeneity of final product. With the mixer and disperser manufacturer's intelligent maintenance system, one can record when additions were made, as well as noting the location of additions, a valuable tool in identifying discrepancies between batches and greatly aiding in product consistency.

A commonly used impeller type is known as an axial pumper (see Figure 11.2). This type of impeller is typically mounted higher up on the shaft and used in concert with a standard impeller design at the base of the shaft. This configuration is useful when mixing highly viscid materials which require additional force to be pushed downward toward the disperser blade, or when a tank is especially large or deep. An axial pumper can help to draw material in from the surface; there are many blade options available, and careful consideration of each will ensure maximally effective results. Holding the impeller is a finely turned and ground shaft. The shafts are commonly between 1 and 5 in in diameter, depending upon the size of the machine, the specific gravity, and the viscosity of product it will disperse.

For thorough dispersion, an impeller should be roughly one-third the diameter of the tank it is mounted in, and roughly one blade diameter from

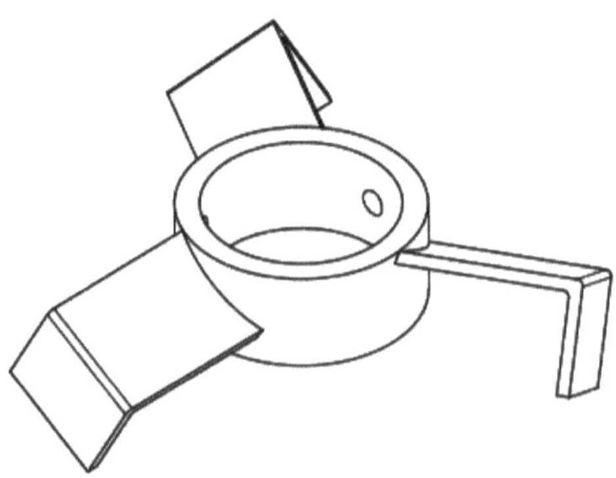

Figure 11.2 Axial pumper impeller. *Courtesy of Myers Engineering, Inc.*

the bottom of the tank. The impeller should not be more than three blade diameters from the top of the batch (depth once all components are charged into tank), and no less than one-half a blade diameter distanced from the top of an uncharged, premixed batch. One must note that these guidelines are generalities, and that impeller sizing and position will vary with the viscosity of the product to be dispersed.

Thc shape of the tank used in mixing will influence the flow characteristics of the product as it is moved by the impeller. Cylindrical tanks are preferred over square or rectangular tanks, as there are no corners which may cause the product to slow down or become entrapped. Square tanks are sometimes used in material let down, when the dispersed material is thinned with a solvent and tinted for preparation in becoming the final product.

Most dispersers are equipped with a tachometer to show the revolutions per minute (RPM) of the shaft. However, for an accurate indication of the force applied to the product, the shaft speed is typically translated into tip speed of the impeller, and denoted in feet per minute (FPM). This is calculated by multiplying the diameter, D, of the blade (in feet) by the RPM of the shaft, and multiplying by π. The practical relevance of this is that a 4-in diameter impeller will have a tip speed of roughly 5000 FPM when the shaft is rotating at 5000 RPM, whereas a 18-in impeller will have a tip speed of 5000 FPM for a shaft rotational rate of 1062 RPM.

Example 1: A process requires a blade tip speed of 5000 FPM for the desired level of dispersion and distribution of all components. What shaft RPM would be needed for a 4-in diameter blade? An 18 in diameter blade?

$$\begin{aligned}\text{Tip Speed (FPM)} &= D \times \pi \times \text{RPM}\\ 5000\ \text{FPM} &= (4\ \text{in})/(12\ \text{in/ft}) \times 3.14 \times \text{RPM}\\ \text{RPM} &= 4777\\ 5000\ \text{FPM} &= (18\ \text{in})/(12\ \text{in/ft}) \times 3.14 \times \text{RPM}\\ \text{RPM} &= 1062\end{aligned}$$

Because the 18-in blade is 4.5 times the diameter of a 4 in blade, the 4-in blade must have a shaft rotational rate 4.5 times greater in order to achieve the same tip speed.

As can be seen by the above example, it is critical that the FPM tip speed be used instead of RPM shaft rate to ensure that the impeller is moving at the correct speed to provide the necessary dispersion.

Another option for measuring the amount of work done by the impeller is the use of a load meter. When used with a disperser, a load meter provides a tremendous amount of data about the rheological characteristics of the product, such as thixotropy or shear thinning, dilatancy or shear thickening, and the viscosity of a product as it is being manufactured, as well as reducing the amount of time necessary to take a sample to a laboratory in order to test the viscosity. As blades age, they become worn down and thinner; a common problem—and cause—of a poorly mixed batch is the use of a worn-out dispersion blade. The blade itself is not the problem, so much as the operator assumption that a worn blade will transfer energy to the product in the same manner as a new blade.

One can envision, for example, your hand stuck out of a moving car window. When your palm is facing the wind, you can feel the force exerted on it; when you drop your palm so it is parallel with the ground the force is reduced. A thinned blade is like your palm turned toward the ground; less force is "felt" by the blade. The force in question is the pumping or dispersing action of the impeller.

Having a load meter helps to track blade wear; when the load is noticed to be decreasing over time, one can mix longer and/or at a higher RPM to achieve the desired dispersion. The majority of individuals in the industry look only at RPM while mixing, but with the use of load cell data in conjunction with RPM, one can maintain more consistent results. In applications where blade wear happens quickly, it is more effective to maintain a consistent load while varying RPM. As the blade wears, the amount of work it does is reduced, so by holding the load constant, you better control the amount of work being applied to the product, because the amount of work being done is directly related to the load.

Mixing and Dispersing Hollow Glass Microsphere Products

The incorporation of hollow glass microspheres into an epoxy system requires different mixing techniques than heavy-filled epoxies which incorporate typical fillers (such as talc, calcium carbonate, fumed silica, etc.). The dispersion of heavy fillers necessitates the use of high-shear impellers which would break the hollow glass microspheres which are intended to lighten an epoxy. In the past, microspheres did not have the crush strength that current technologies do, and were much more susceptible to breakage in traditional mixing and dispersing applications. However, hollow glass microspheres must still be incorporated with care, as they are more fragile than standard fillers, and therefore are susceptible to damage through the use of high shear. With advances in technology, dispersion equipment manufacturers and suppliers have been able to develop machines which can aggressively disperse these dense fillers as well as gently mix hollow glass microspheres—with minimal glass microsphere damage—at a variety of product viscosities.

Radial Pumper ("Super Pumper")

Various impeller designs and configurations are used, depending on the batch viscosity and processing requirements for components within the batch. Unfortunately, despite all the technologies currently available, most companies—regardless of how big or small—do not properly apply the different types of dispersion. More than 90% of people within the industry believe that a proper tip speed for a disperser is 5000 FPM. While this speed is required for particle size reduction and deagglomeration, it is by no means necessary or efficient when used in blending applications, or in applications that contain shear-sensitive components. Blade selection is also not well understood, and there are types of blades that should be used for one type of process yet not another. For glass microsphere-filled products, it is essential to minimize shear forces, which can be damaging to the hollow glass microspheres and ultimately lead to an increase in product density from theoretical ideals.

The radial pumper impeller (see Figure 11.3) is an exemplary choice when incorporating microspheres into thixotropic and/or viscid materials. The basic principle behind the radial pumper impeller is that the angled teeth sling the material outward, continually accelerating the perpetual batch movement; the teeth do not shear the product.

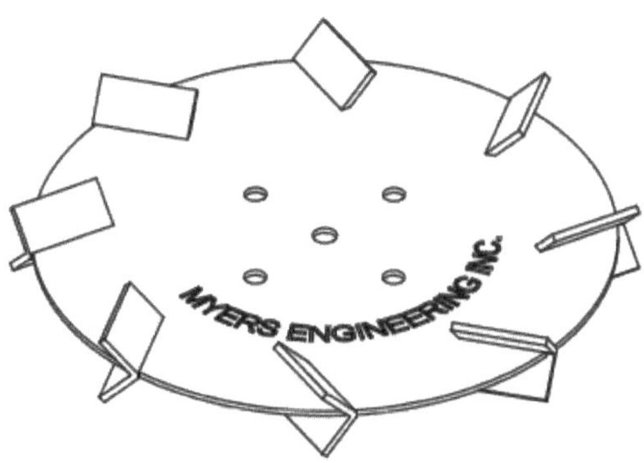

Figure 11.3 Radial pumper impeller design. Forward facing teeth accelerate product with minimal impact shear. *Courtesty of Myers Engineering, Inc.*

By creating perpetual batch movement, viscid material (from 10,000 to 150,000 cP and above) can be thoroughly mixed while maintaining a low-shear environment. Forward facing teeth provide longer contact between the impeller and product captured in the flute; this results in a higher velocity or product energy each time the hollow glass microspheres are in contact with the impeller surface. The hollow glass microspheres remain in contact with the impeller surface longer, developing a higher velocity and energy transfer, which enables the mixer to run slower, resulting in less initial impact force upon the hollow glass microspheres. This blade configuration allows for a rolling vortex to be achieved without the damaging forces of traditional high speed dispersion.

The radial pumper physically pushes the product outward, creating an area of low pressure underneath the shaft, as product rushes back to fill in the void in an attempt to equalize pressure and product distribution. This low-pressure void is a good place to add hollow glass microspheres into the batch, especially when the mixing tank being utilized does not have a good seal. When hollow glass microspheres are incorporated subsurface in this manner, they are thoroughly assimilated into the liquid product. This enables rapid blending without exposing hollow glass microspheres to the atmosphere, a potential health and safety hazard. The material within the tank acts as a filter, by wetting out the hollow glass microspheres so they cannot become airborne.

A common complication caused by addition of hollow glass microspheres atop the batch is that the initial charge of hollow glass microspheres become wetted (i.e., are absorbed into the surface), while subsequent amounts of hollow glass microspheres "flour" the surface of the batch. Once the surface becomes fully saturated with a dry substance such as hollow glass microspheres, it can no longer draw in new material. This is why one should add hollow glass microspheres at the vortex, so they are quickly drawn into the product and do not sit on the top of the batch, preventing incorporation of later additions of hollow glass microspheres, or leading to agglomeration.

Their microscopic size, along with their low aspect ratio, leads to the hollow glass microspheres behaving as ball bearings upon the top of the batch, effectively rubbing against one another while doing little in the vein of incorporation and assimilation. Due to their low density, the hollow glass microspheres are not readily pulled downward into the batch by their own mass; and due to their sensitivity to shear, one cannot utilize mixing speeds which would easily incorporate denser fillers such as talc, calcium carbonate, or glass fibers. To combat this dilemma, one should integrate hollow glass microspheres in the vortex created by the dispersion blade atop the product; or if subsurface addition is preferential, incorporate hollow glass microspheres at the low-pressure void created directly under the blade and shaft assembly. Wherever one chooses to incorporate hollow glass microspheres, it should be at a location such that the hollow glass microspheres can be directly injected into the region of laminar flow (i.e., not an area of low or no velocity).

High-Shear Disperser Blade

High-shear dispersion blades (Figure 11.4) can be used when good dispersion and deagglomeration of dense fillers and pigments is needed, with a variety of product viscosities, and when shearing of the product is required. A typical periphery speed for a high-shear dispersion impeller is 5000 FPM; this speed causes damage to hollow glass microspheres, but is required to properly break down other fillers within the batch. The high-shear dispersion blade utilizes the principles of point shear and pumping to mix higher viscosity products.

Pumping is the physical act of the impeller moving product aside as the impeller rotates. This process centrifuges the product outward in a radial direction until it reaches the wall of the tank, at which point the flow is split horizontally into two; this forces the flow of material up and down the side wall of the tank. The upward flow of material "wells" up the side of the tank to the surface where it loses velocity, and then rolls back to the center of the vortex where it then falls to the disperser blade to begin the process all over again. The downward flow continues until it hits the bottom of the tank, where it then changes flow direction—picking up any undispersed solids that have settled or stuck to the bottom of the

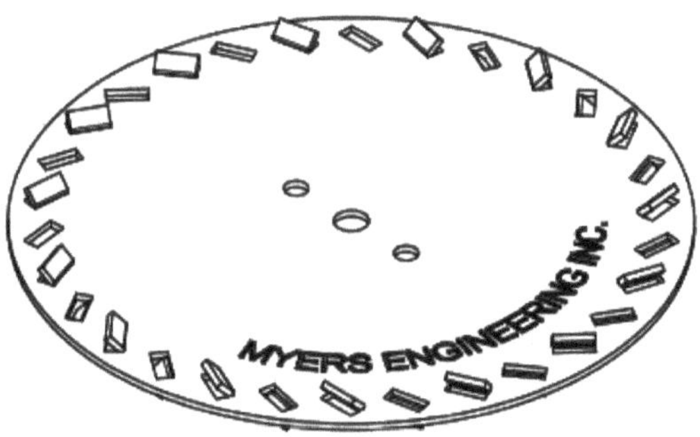

Figure 11.4 An illustrative high-shear dispersing blade. Various mixing, dispersing, and flow characteristics can be obtained by augmenting the number, size, and location of "teeth." *Courtesy of Myers Engineering, Inc.*

vessel—while moving to the low-pressure zone underneath the dispersion blade. The material is then drawn upward where it is reintroduced to shear, thus continuing this recursive process. The key to this configuration is perpetual batch turnover, which continually reintroduces the particles to the high-shear zone.

Mixing Dynamics and Dispersion Blade Placement

Proper blade placement is critical to obtaining good dispersion and a homogenized mixture of ingredients. The ideal distance from the bottom of the mixing tank to the blade is typically between ½ and 1 times the blade diameter, depending on the substance being mixed, its viscosity, and the tank diameter to liquid height ratio.

The further down the blade is within the tank, the more work the blade must do to move product at the top of the batch. This will lead to destruction of hollow glass microspheres, as they are being ground at the bottom of the batch—where the rapidly rotating dispersion blade is creating high shear—and not fully being incorporated at the top of the batch. A limited-lift style mixer can remedy this issue; as the shaft can be raised or lowered, the position of the impeller—and hence the location of the vortex—can be altered. The lack of mixing at the top of the batch can be due to many things, such as a viscous fluid which does not readily transmit energy and/or movement, or due to the lightweight nature of hollow glass microspheres not sinking into the batch by their own gravity.

In utilizing a lift style mixer, the dispersion blade can be raised closer to the surface of the batch; this maintains a vortex for efficient addition of hollow glass microspheres while putting less energy into the product, and thereby minimizing glass microsphere damage. Once a certain volume percentage (approximately 30% by volume) of hollow glass microspheres is added, the specific gravity of the product is reduced to a point where the vortex is reduced or completely eliminated. Placing the dispersion blade at the proper location and depth for a specific application and choosing a blade with an optimal power number for said application will minimize mix time, glass microsphere damage, and energy consumption. See Figure 11.5 for an example of a standard tank and blade configuration.

Mixing of viscous systems includes two main mechanisms: dispersive and distributive mixing. The goal of dispersive mixing is deagglomeration, or attrition of particle size. The purpose of distributive mixing is to produce a homogeneous compound, where particles are evenly distributed

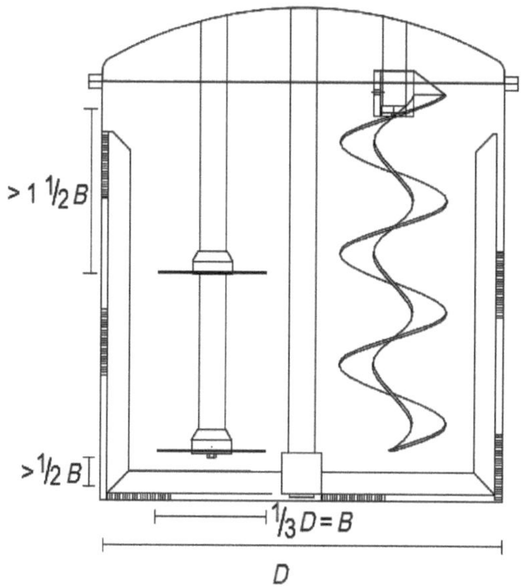

Figure 11.5 Standard ratios used in sizing a disperser.

in the resin matrix, as the nomenclature implies. Dispersive mixing is an intensive process, aimed at reduction in agglomerates. Distributive mixing is an extensive process focused on spatial uniformity. When one thinks of mixing and dispersing equipment, generally both distributive and dispersive mixing are required to achieve the desired output. Figure 11.6 illustrates each type of mixing, and the end result when the two processes are combined.

Dispersive and distributive mixing are the main means by which complex solutions of solids and viscous liquids are formed. Dispersion is the process of breaking down aggregates and/or agglomerates down to their elementary particles. Aggregates are clumps of particles which are primarily joined at their faces, forming a mass which has significantly less surface area per volume than the base particles. Agglomerates, on the other hand, are joined at their edges and corners, resulting in a volume that has nearly the same surface area as the combined surface area of constituent particles. Based on this geometrical difference in structure, aggregates are much more difficult to break up than agglomerates; aggregates are much stronger due to the decreased surface area of the solid, in conjunction with greater intermolecular forces.

For hollow glass microspheres, a blade with higher product displacement than product dispersion is ideal. For example, a radial pumper impeller will displace the product while reducing shear, as compared to

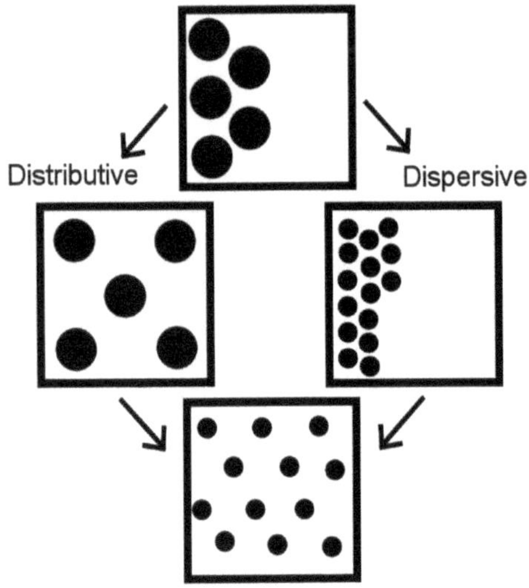

Figure 11.6 Schematic representation of distributive and dispersive mixing processes. The two modes combined result in a homogenized product with agglomerates reduced to primary particle size.

a high-shear dispersion blade. Product displacement is related to tip speed of the blade; however different impeller configurations will transmit energy differently, and necessitate varying shaft rotational rates and tip speeds. For example, aluminum paints should be mixed at a tip speed of 1200–1800 FPM; this ensures thorough mixing, while minimizing destruction of aluminum flakes. An analogy can be drawn between aluminum paints and glass microsphere-filled liquids and pastes; in that a homogeneous mixture is desired yet there are materials in the product that may be damaged if a high tip speed is used during mixing. For such an application, a radial pumper impeller is desirable, as it moves the batch and provides adequate homogenization while reducing shear forces, in comparison to a high-shear disperser blade.

Mixer Design Options

A variety of single, dual, tri, and quad shaft mixers are available for a multitude of applications and product viscosities. Single shaft radial pumper impeller mixers are best suited to low-viscosity glass microsphere products, such as paints and coatings. High viscosity mixtures (such as caulking compounds and putties) are thoroughly mixed with minimal glass microsphere damage, utilizing dual- or tri-shaft styles of mixers.

Single Shaft Dispersers

Single shaft mixers and dispersers utilize one of a variety of dispersion and mixing blades to quickly and efficiently combine materials for a uniformly dispersed product. A standard disc impeller blade (see Figure 11.4) creates laminar flow to disperse material beyond the edge of the blade tip for particle size reduction. The alternating teeth create high shear radially about the impeller; this is where dispersive mixing takes place.

Dual- and Tri-Shaft Dispersers

Multishaft dispersers combine multiple blades, augers, and/or sweepers within the confines of a single tank to maximize usefulness, productivity, and reduce multiple handlings of a batch to obtain a homogenized end product. Dual- and tri-shaft mixer designs typically include one impeller blade, with the other shaft(s) comprised of a helical auger, a helical (ribbon) sweeper, and/or sweepers with fixed or articulated wipers.

Multishaft dispersers incorporate multiple blades, using different speeds and flow patterns so that each can develop maximum results at each phase of the batch. A mixer and disperser manufacturer produces mixers which use two to four different speed ranges with an appropriate blade to maximize mixer performance, minimize batch time, cleanup, and waste. As discussed in the basics of dispersion, most products start with a high speed dispersion impeller to disperse on a fine level, much like a blender.

As the complexity of the product increases through variations in viscosity, temperature requirements, specific gravity, evacuation of air, pressurization, and the sensitivity of the raw material being added, the following additional shafts may be incorporated in the design of a mixer.

The low-speed shaft is a typical secondary shaft that is added to the representative disperser. In bygone days this was done to help material flow as the viscosity and specific gravity became too great for a single shaft. In subsequent years, the tanks became jacketed for heating and cooling. When this was done there was a problem in that the material that was stuck to the side of the tank would become an insulator, thus preventing the remainder of the batch from necessary changes in temperature. At this point in time, risers and wipers have been added to the sweep blade of the low-speed shaft. Functioning much like a spatula, the wiper removes the material from the wall of the tank, allowing fresh material to contact the wall of the tank for adequate heat exchange and thorough distributive mixing.

Sweep blades come in several types. The "low profile" blade is primarily used for the sole purpose of wiping the walls of the tank. Due to this blade's low profile design, it does not move much material and therefore does not require a great deal of horsepower, keeping the capital cost—as well as operating cost—of the equipment down. Conventional peripheral speed for a blade of this type is on the order of 400 FPM.

The next step up is a "pitched angle" blade. This blade is slightly stiffer than the previous. The pitch of the riser gives this blade the ability to provide a greater displacement, thus moving more products forward. As the product becomes more viscous, this forward movement of the mixture ensures that the high-shear zone is perpetually fed with new material. Tip speed range of this blade is 400–600 FPM.

Next we have the "box angle" blade. Shaped like a triangle, this blade is extremely strong and is used when the considerable viscosity could cause a typical sweep blade to twist. This blade also has a large displacement and will move product more aggressively than a low profile design.

The final low-speed blade is a helical sweep. This blade has the strength of a box angle blade, while incorporating a helical flight which runs the perimeter of the tank. The purpose of this blade is to assist in lifting the material up the side of the tank and bringing it to the surface of the batch. As the material moves to the surface, it creates a void at the bottom of the batch, while simultaneously causing the material from the center of the mixer to be moved out to the wall of the tank, where it will be exposed to heat transfer and be lifted to the surface, completing one iteration of the this cyclical process. This blade is a superlative component in adding low-viscosity—or low density—materials. In the case of hollow glass microspheres, the gentile folding action will work the hollow glass microspheres into the batch and mix them evenly from top to bottom. When mixing hollow glass microspheres, mixers can utilize a dual-direction blade to engage the wipers only when wiping is absolutely essential, as the close tolerances essential to effective wiper action will grind the hollow glass microspheres, thereby destroying them.

A helical—or ribbon—sweeper (see Figure 11.7) functions similarly to a sweeper with wipers, as described above, yet with the added benefit of overturning the product as it rotates. The ribbon sweeper lifts the product upwards as it rotates, which thereby causes the material to fall back down upon itself in the center of the batch. This is a useful shaft to incorporate into the design of a glass microsphere-product mixer, as it works the hollow glass microspheres into the solution without the damaging effects of high energy.

Wiper blades are typically made of nylon, micarta, or teflon. Wipers are effective for homogenizing a mixture, as well as efficiently transferring heat away from the shaft and outward toward the outer wall of the mixing tank. However, wiper blades, by their intent and nature, have extremely

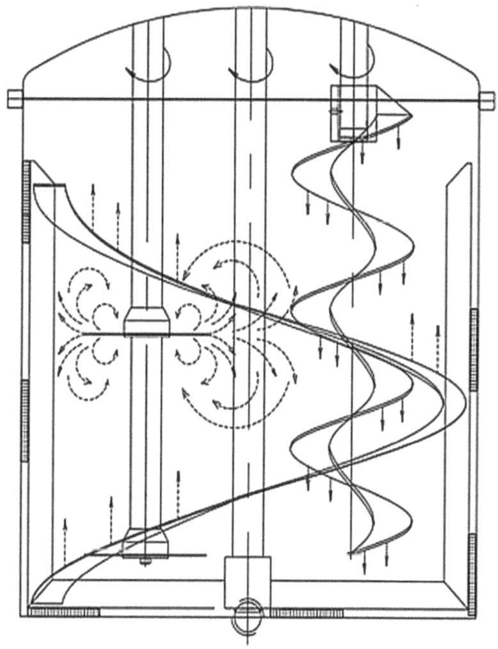

Figure 11.7 Typical blade configuration of a multishaft mixer. Flow lines illustrate the various functions served by each blade. Blade type, front left-to-right shaft: twin high speed disperser blades, low-speed sweep with articulated wipers, and intermediate speed helical auger.

close contact with the sides of the tank, and may cause hollow glass microspheres to be crushed or damaged as they move product away from the wall of the tank. To minimize glass microsphere casualty, an articulating wiper blade sweeper should be utilized. As an illustrative case, the sweeper can rotate clockwise (employing the sweeper wipers) when hollow glass microspheres are not yet incorporated, and good heat transfer is desired. Once hollow glass microspheres are added to the product, the sweeper is then made to rotate counterclockwise, disengaging the wiping action of the articulated wipers; this allows for good mixing while ensuring that hollow glass microspheres are not damaged in the process of their addition to, and incorporation within, the batch. The sweep blade still rotates to provide distribution, but without the wipers engaged there is ample clearance so as to not damage the hollow glass microspheres.

The selection of intermediate speed blades comes in two ways: first, a reduced speed disperser, known as a blender. Running between 3000 and 4200 FPM, this range will provide good homogeneity during mixing. As a result of the reduced speed range, one can get more work done with a lower horsepower, saving limited energy resources.

The bulk of the intermediate speed impellers are in the 800–1400 FPM range. One example of this is a gate blade, which functions similarly to an egg beater. This blade will work the material by stretching it much like a taffy mixer. This blade is not effective when the product becomes high in viscosity or when the specific gravity falls below 0.8. When this happens, the gate traps a ball of material in the center of the blade which will not be subjected to any vertical movement. By itself, the gate only works the material, and does not offer any vertical benefit.

Turbine blades will offer some vertical movement, as well as some working of the material in the impact zone. A turbine blade offers better mixing than a gate blade, while being less aggressive than a dispersion impeller. For this reason, in the last couple of decades, this has been a common method for incorporation of hollow glass microspheres.

Next is an auger, which typically has a pipe in the center of the flight for support. While this is a preferred design for some products with high viscosity—due to the design which provides extra strength—it is hard to clean and will "pack" with material. To understand the packing concept, visualize the accumulation of mud or snow in the tread of a tire. The auger can become filled with a ball of material, inhibiting proper function.

A helical auger is similar to the auger, but does not have the pipe in the center (see right-hand side of Figure 11.7). Strength lost by the removal of the pipe in the center is made up for by building the flights out of heavier material. With an open center, this blade is much easier to clean. As the blade surface comes in contact with the material, it presses the material down into the batch. As this happens, some material slips to the outside diameter (OD) and inside diameter (ID) of the flight. Unlike the full auger, the material that slips to the ID hits the center of the auger where it is trapped and cannot get out without being pressed to the bottom of the tank. The cavity that is created in the center of the helical auger will rapidly fill with hollow glass microspheres; because the hollow glass microspheres cannot escape the center of the blade without mixing material, this blade is the absolute best option for highly-filled or higher viscosity material that contains hollow glass microspheres. This blade also offers a great advantage for de-gassing of material. When the blade is run in reverse, it pulls the material up from the bottom of the mixer and allows it to fold out on the surface. This action of folding out (as opposed to "folding in") reduces the surface tension of the material and makes it easier to release the micro-sized air bubbles. Application of vacuum expedites this process, as larger bubbles of air can be transmitted to the top of the batch and air removed more rapidly.

Aside from the upward drive of the ribbon sweep, dual- and tri-shaft mixers can be equipped with an auger which delicately drives the hollow glass microspheres downward. The helical auger gently opens a void at the top of the product batch which hollow glass microspheres can fall into, being deeply incorporated into the mix, and minimizing the problem of the hollow glass microspheres "flouring" the top of the batch. The hollow glass microspheres cannot escape, thus they are folded and worked into the mixture; this creates good glass microsphere integration with the rest of the material. Figure 11.7 shows a multishaft mixer which incorporates dual high-shear dispersion blades on a single shaft, a ribbon sweep, a helical auger, and scrapers. The advantage of a multishaft disperser is the generation of a variety of streamline and flow patterns within a single tank. For this reason, improved homogenization can be achieved in a shorter time frame, with less equipment, and less product handling.

The versatility of a tri-shaft design allows one to accomplish thorough dispersion of polymers and additives, and then shut down the high-shear impeller when ready to add hollow glass microspheres. For high viscosity applications, this system is many times more efficient and effective than any other mixer; other systems may require moving a batch one or more times to obtain a homogeneous mix of all components of the product. Henceforth, a dual- or tri-shaft mixer can eliminate the need for double handling of the material. For example, in the past, products have been compounded by creating the dispersion base on a dissolver, then transferred over to a sigma (or paddle) style mixer. This reduction in handling and number of steps to create a batch has lead to an increase in output and a decrease in square footage demand on a mixing plant, increasing the efficiency, productivity, and profitability of a compounding facility.

Another mixing option which can improve disperser safety is to use a sealed—or enclosed— mixer, thereby reducing operator exposure to glass microsphere inhalation. All mixers can be either atmospheric (i.e., without a lid, or, with a lid that does not change the pressure within the vessel), or sealed, and thus capable of having vacuum applied. A sealed tank allows for inerting the atmosphere within the tank (e.g., with nitrogen or other nonreactive gases) and reducing the pressure within the tank to facilitate batch production. Pulling vacuum on a sealed tank is an efficient means of transporting hollow glass microspheres into a product, either subsurface or at the top of the batch. With hollow glass microspheres especially, it is important to direct their flow into the vortex or void created at the top of the tank; this minimizes "floating" and potential agglomeration. A sealed tank can also be used to raise the pressure, which can be useful in aiding the feeding of pumps.

It is important to pump the glass microspheres in at a rate which they can be incorporated into the batch, otherwise overfeeding of the batch may result. Overfeeding the mixer can lead to the necessity of longer mix time, and consequently, the damaging effects of mixing action on hollow glass microspheres. To this end, it is very important to add hollow glass microspheres quickly and efficiently to minimize shear forces they experience. In general, a good practice is to add hollow glass microspheres toward the end of batch agitation to prevent prolonged exposure to the mixing process. In summary, add hollow glass microspheres once dispersion is complete, obtain a homogeneous mixture with the hollow glass microspheres as quickly as possible, and discharge the batch.

Certain styles of mixers and dispersers should be avoided when hollow glass microspheres are to be incorporated into a product. These include, but are not limited to, planetary, double planetary, and horizontal mixers. Planetary and double planetary mixer designs rely on close contact between disperser blades to thoroughly mix the product by shearing it. This can be problematic, as it may result in significant glass microsphere damage due to the tight clearance of blades rotating opposite of one another, which creates high shear. Planetary and double planetary mixers tend to mix the product in a horizontal fashion, meaning that there is primarily mixing and dispersion in the lateral direction. This depth-striated mode of mixing does not quickly incorporate hollow glass microspheres, contributing to a longer mix time, and an increased risk of glass microsphere damage. Horizontal mixers use a more gentle folding over of the product, but this style of mixer can be difficult to clean, and the product can leak out of wearing seals at the ends of the mixer.

Difficulty in Mixing of Hollow Glass Microsphere-Filled Products

When mixing liquids of vastly differing viscosity a problem arises, in that the mixing will take longer than with liquids of similar viscosity, as well as the necessity of more energy being applied to achieve a homogenized dispersion. An analogy can be drawn between a viscous solution of resins and hollow glass microspheres. Hollow glass microspheres, although a solid, exhibit flow characteristics similar to inviscid liquids, particularly when fluidized, as during transport to a mixing vessel. If one considers the fluidized hollow glass microspheres the "thin" liquid, and the resin or composite matrix as the "thick" liquid, it can be seen that mixing will not readily take place. In addition to this correlation, it is noted that the mixing of floating solids (i.e., lower in density than the liquid they are added to) requires greater energy input than the mixing of neutrally buoyant or sinking solids [1].

Current literature and research is lacking in the area of dispersion of floating solids, especially solids of small size. For a more thorough understanding of mixing of hollow glass microspheres, additional advances in the area of their specific viscosity, rheology, and flow and dispersion characteristics in matrices of moderate to high viscosity must be investigated.

As stated by Atiemo-Obeng et al. [2], "solids that float without agitation include solids that are less dense that the liquid, dense solids with trapped gas, and solids that are difficult to wet." [2]. hollow glass microspheres exhibit all three of these dispersion difficulties: they are composed of soda-lime borosilicate glass which is more dense than the matrix they are combined with, yet due to their "shell thickness" and entrapped gases, are actually less dense than the matrix; being made of glass, which is a highly non-porous substance, they do not truly "wet" in the same sense that a clay, talc, or calcium carbonate would wet; and are therefore difficult to wet and incorporate into a product.

In the incorporation of solids, more than just the chemical and physical properties of a material is of importance. Atiemo-Obeng et al. [2] state that "both average size and the particle size distribution are important properties." Hollow glass microspheres come in a multitude of different average sizes and size range variance. This ensures that the size of an average glass microsphere, as well as the complete scope of diameters encompassed, will be suitably incorporated in the dispersion process, with minimal breakage. The greater the size distribution, the more difficult it can be to homogenize a mixture. This is due to the smaller particles filling voids between larger particles, effectively creating agglomerates and regions which entrap air and do not allow the matrix material to fully penetrate.

Effect of Particle Shape and Size on Dispersibility

A multitude of factors can impact the ease or difficulty with which one material may be combined with another. Research by Atiemo-Obeng et al. [2] shows that large, dense particles are harder to suspend in solution than light particles; spherical particles–due to their aspect ratio of unity—are harder to suspend than particles of high aspect ratio, such as platey talcs or glass fibers. One must weigh what factors contribute more or less to a particle's incorporation into the continuous matrix. It can be assumed that a small, non-dense, spherical particle will also be difficult to incorporate, yet for nearly opposite reasons than a large, dense particle of virtually any shape or aspect ratio. This need to hypothesize or draw conclusions based on an inference from dissimilar fillers reinforces the necessity for additional research to be performed on the rheological properties of, and dispersion of, hollow glass microspheres in solution.

Location of Additions

Dispersional studies by Weetman [3] confirmed the importance of location of additions to a batch: "typical low velocity regions occur near the top of the tank (where feeds are often introduced) and underneath a large pitched blade impeller (if used) at the bottom of the tank." [3] This suggests that additions of hollow glass microspheres at the top of the batch may not be fully incorporated, evenly dispersed, or may require a longer duration of mixing than anticipated to become thoroughly incorporated. To expedite the process, it is recommended that hollow glass microspheres be charged into the batch as close to the vortex as possible. This ensures that—due to their low density, as well as flow characteristics—they do not "flour" the top of the batch, which would lead to agglomeration and/or prolonged mix time (which may result in extensive glass microsphere breakage).

The effect of orientation of additions into the matrix is apparent in solutions which are viscous, non-Newtonian, and exhibit laminar flow [4]. An initial orientation of concentric addition will result in a concentric distribution after many rotations, that is, the hollow glass microspheres will not be evenly distributed throughout the batch, but striated within a fix radius about the tank. This is akin to additions which are "point focused" (see Figure 11.8). Conversely, additions which are made in a radial manner will more quickly disperse throughout the entire batch (Figure 11.9), and may be thought of as a "line" addition, as opposed to a "point" addition. By the simple nature of the geometrical terms used to describe these two methods of additions in a laminar flow regime, one can readily grasp that concentric addition is essentially one dimensional, whereas radial addition exemplifies a two dimensional means of incorporation. From this simple analogy one can easily infer that a two dimensional addition will result in homogenization more expediently than a one dimensional addition. The radial addition has the added benefit of hollow glass microspheres immediately being pulled into the vortex, by broadcasting the addition across the batch, toward the center of the vortex.

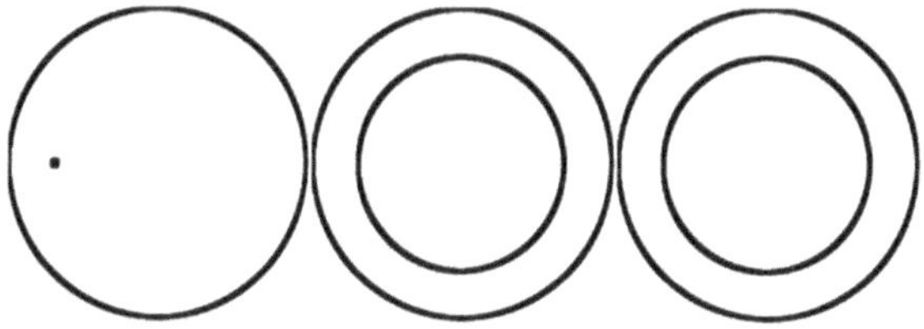

Figure 11.8 Concentric addition at instant of addition, after one rotation, and after many rotations.

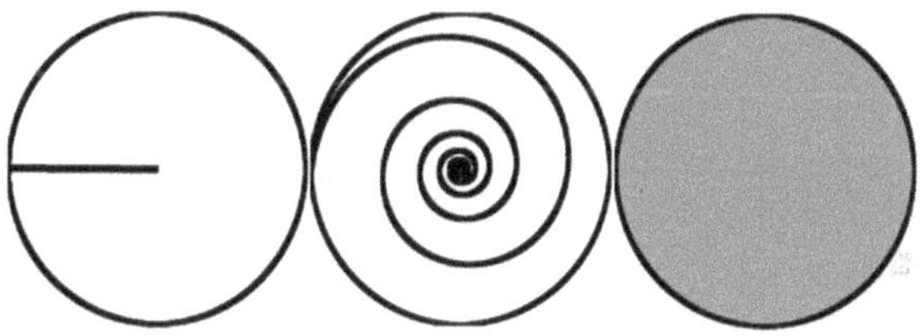

Figure 11.9 Radial addition at instant of addition, after a few rotations, and after many rotations.

The Science of Mixing and Dispersing

Surface Tension

The ease with which a solid may be dispersed within a liquid is a function of adhesion tension, which is often considered to be an indication of the degree of wetting. Adhesion tension is a function of contact angle (between a solid and a liquid), surface tension, and the interfacial tension between the solid and liquid [5]. The Young–Dupre equation exemplifies this relation.

$$\gamma_{SA} - \gamma_{SL} = A_{SL} = \gamma_{LA} \cos \theta \qquad (11.1)$$

where

γ_{SA} = "surface tension" of the solid

γ_{SL} = interfacial tension of the solid and liquid

γ_{LA} = surface tension of the liquid

$A_{SL} = \gamma_{SA} - \gamma_{SL}$ = adhesional tension

θ = contact angle

The above equation conveys the interplay between these physical phenomena.

Wetting Phenomena

Wetting phenomena are directly related to the surface tension of the liquid, γ_{LA}, and the liquid–solid contact angle, θ. As a general rule, when γ_{LA} is greater and/or θ is smaller, wetting, and hence mixing, occurs more rapidly.

Figure 11.10 shows various important contact angles for a liquid–solid system. When the contact angle, θ, is between 90° and 180°, adhesional wetting takes place and is spontaneous. When the angle of contact is less than 90°, immersional wetting (i.e., immersion of the solid phase into the liquid phase) takes place (conversely, when $\theta > 90°$, immersional wetting is not spontaneous, and the solid can be considered to be emerging out of the liquid phase). Spreading wetting occurs only when $\theta = 0°$, and the solid is preferentially transferred into the liquid. When $\theta = 0°$, a substance is considered to be hydrophilic; for substances or solids which do not experience a 0° contact angle, varying degrees of hydrophobicity are considered to exist.

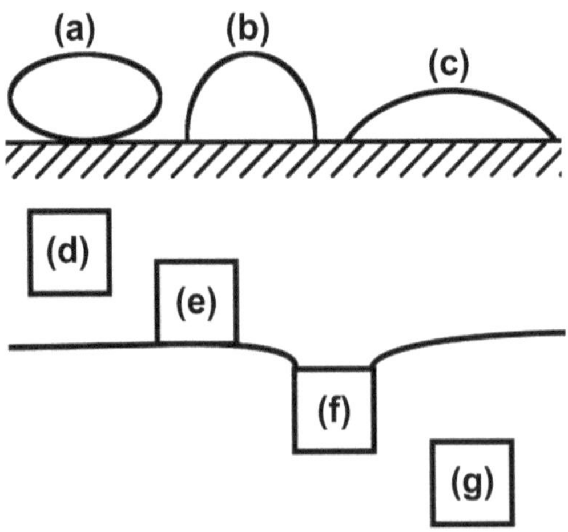

Figure 11.10 Important contact angles in the mixing process: (a) adhesional wetting, contact angle $90° < \theta < 180°$; (b) immersional wetting, contact angle $\theta = 90°$; (c) spreading wetting, contact angle $\theta < 90°$; (d–e) adhesional wetting; (e–f) immersional wetting; and (f–g) spreading wetting.

Estimating Effects on Viscosity

As solids are added to a liquid, or preexisting solution of solids and liquids, the viscosity changes—the new solution tends to be more viscous. The viscosity of a slurry can be determined from the viscosity of the liquid (or liquids, or solution), and a dimensionless parameter known as the relative viscosity, which takes into account the volume fraction of solid particles contained in the slurry, as follows:

$$\mu_s = \mu_r \mu_l \tag{11.2}$$

where

μ_s = slurry viscosity

μ_r = relative viscosity

μ_l = liquid viscosity

The slurry viscosity is proportional to the liquid viscosity, with the constant of proportionality being the relative viscosity, which is dependent upon many factors, that is, filler loading, filler shape (aspect ratio), and smoothness.

Einstein developed an equation to determine the relative viscosity, however, the equation is only valid at low loading concentrations, and for very fine particles:

$$\mu_r = 1 + 2.5\phi \tag{11.3}$$

where ϕ = volume fraction of solid particles.

Guth and Simha [6] modified this equation to consider particle–particle interactions at higher filler loadings:

$$\mu_r = 1 + 2.5\phi + 14.1\phi^2 \tag{11.4}$$

Thomas [7] modified this equation further by fitting with empirical data:

$$\mu_r = 1 + 2.5\phi + 10.05\phi^2 + Ae^{B\phi} \tag{11.5}$$

where

$A = 0.00273$

$B = 16.6$

The above equations are for use with any filler shape (and within a wide range of filler size distributions). Further work by Kitano et al. [8] modeled the basic Einstein equation for use with very high loading concentrations of particles:

$$\mu_r = (1 - \phi/A)^{-2} \tag{11.6}$$

where A = 0.68, for smooth, spherical particles.

The Kitano et al. equation is based on their work done with glass microspheres, glass balloons, and silas balloons in viscous particle-filled polymer melts [9]. Modified forms of this equation are:

$$\mu_r = (1 - \phi/\phi_0)^{-2} \tag{11.7}$$

$$\mu_r = \exp[(2.5\phi)/(1 - k\phi)] \tag{11.8}$$

where

$\phi_0 = 0.68$

$k = 1.25$

are constants relating to the packing factor of spherical particles with a Gaussian size distribution within less than one standard deviation. Kitano et al. [9] found that these equations are useful in predicting μ_r versus ϕ for $\phi \leqq 0.4$, and are better fitted for high volume loadings than the equation derived by Thomas. The reason for this better fit is due to their simplicity, and the inclusion of only one constant, which is a function of packing factor of the filler, as opposed to empirically derived constants.

To illustrate how the modifications to the Einstein equation impact the value of the slurry viscosity, let us use an example.

Example 2: Assume a liquid has an arbitrary viscosity equal to 1.0. Using Eqns (11.2)–(11.8) to determine the relative viscosity ratio, what would the slurry viscosity be if 30% by volume ($\phi = 0.3$) microspheres is added to the liquid?

$$\mu_s = \mu_r\mu_l = 1 \tag{11.2}$$

$$\mu_r = 1 + 2.5\phi = 1.75 \tag{11.3}$$

$$\mu_r = 1 + 2.5\phi + 14.1\phi^2 = 3.019 \tag{11.4}$$

$$\mu_r = 1 + 2.5\phi + 10.05\phi^2 + Ae^{B\phi} = 3.0516 \tag{11.5}$$

$$\mu_r = (1 - \phi/A)^{-2} = 3.2022 \tag{11.6}$$

$$\mu_r = (1 - \phi/\phi_0)^{-2} = 3.2022 \quad (11.7)$$

$$\mu_r = \exp[(2.5\phi)/(1 - k\phi)] = 3.3201 \quad (11.8)$$

This is clearly a vast range in values, from 175% to 332% of the liquid's viscosity. We can see through this that as these equations introduce constants related to particle–particle interactions, observations based on empirical data, filler shape, surface roughness, and packing factor that the results seem to converge. The wide array of values indicates that many physical phenomena are at play here, and must be carefully considered in order to predict rheological characteristics essential for accurate sizing of dispersion equipment.

See Figure 11.11 for a schematic representation of how the introduction of various physical phenomena alter (and improve) the prediction of slurry viscosity based on volumetric loading.

Kitano et al. [9] also found that highly filled materials tend to exhibit a yield stress, σ_y, and that yield stress increases with increased filler concentration. The concentration of filler (by volume) and yield stress are related and—when plotted on a log–log scale—produce a linear plot. Kitano concludes that the only appreciable influence a filler has on relative viscosity—at a given concentration—is the size distribution of particles. Hollow glass microspheres are manufactured in an array of sizes, with each class of bubbles being of a narrow size range; this limits variability in product quality and provides optimally measurable and predictable rheological data.

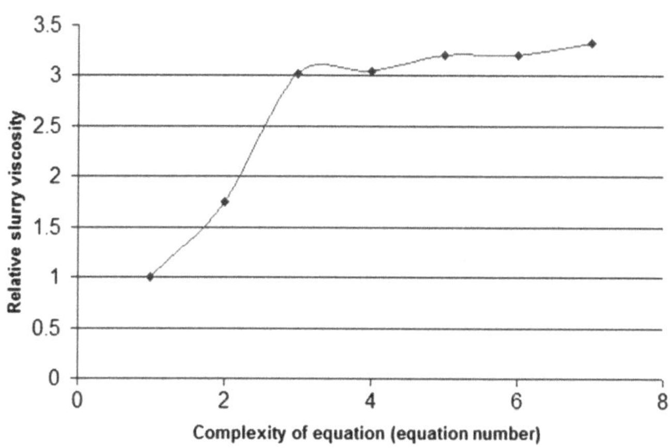

Figure 11.11 Graphical representation of Example 2 showing logarithmic progression in accuracy of slurry viscosity equations.

Rheological Effects on Dispersion

Deagglomeration in viscous solutions is primarily achieved by means of extensional (elongational) flow and (laminar) shear flow. Shear flow is defined as the fluid velocity gradient being at right angles to the direction of flow; this can be thought of as flow through a pipe (see Figure 11.12). Extensional or elongational flow occurs when the velocity gradient is parallel to the direction of flow, as in a contraction, or narrowing region, which the flow must pass through. See Figure 11.13 for visual of these types of flow. Laminar shear flow deforms fluid elements due to the relative motion (velocity) between streamlines. Elongational flow causes the material to be stretched, therefore accelerating it at different rates. Both of these processes result in increased interfacial area between the solids and the liquid(s), and therefore result in reduced striation thickness of additions, a key requirement in providing homogeneous mixture quality. Laminar shear flow "tears apart" the liquid (and any agglomerates within), resulting in a mono dispersed product. These mixing phenomena are primarily responsible for reduction of agglomerates when the solution is viscid, and molecular diffusion is negligible.

When shearing rates are equal, elongational (extensional) shear does more to disperse solids than rotational shear. This is true for a wide variety of viscosity ratios, as well as when the viscosity ratio is near unity.

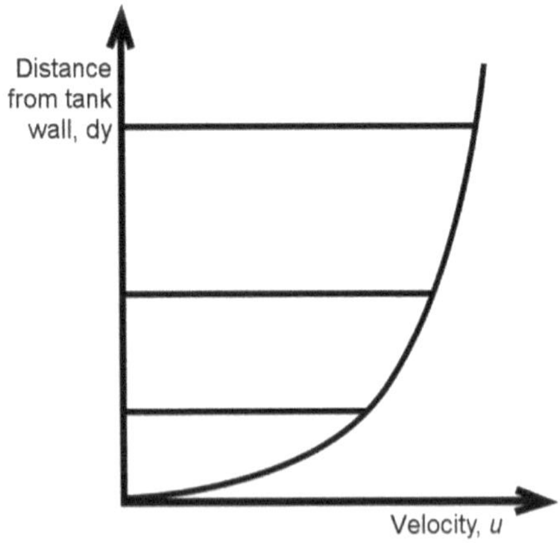

Figure 11.12 Flow rate generated by shear stress. Flow velocity is greater farther from vessel walls.

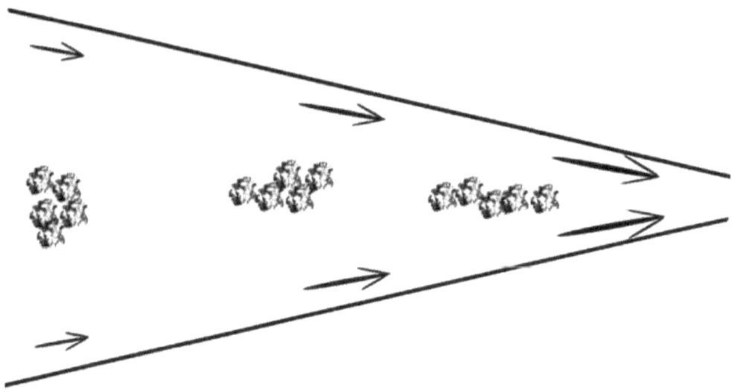

Figure 11.13 Elongational (extensional) flow. Arrows indicate flow velocity.

Compare Figures 11.8 and 11.3, elongational flow and concentric additions (which is evocative of rotational shear forces), respectively. One can envision elongational flow being akin to radial additions; the addition is "stretched" thinner and thinner (i.e., elongated). By this simple analogy it can be more clearly seen that elongational shear is critical to batch, and therefore product, homogeneity.

The rheological characteristics of hollow glass microsphere dispersions tends to be complex, due to the physical attributes of the components involved. The main physical properties to consider are Brownian (dispersive) motion, viscosity, density, buoyancy forces, Stokes (frictional) forces, and the gravitational force acting on the materials. As summarized by Walter, "high viscosities, low differences in density between dispersed particle and dispersant and, especially, small particle sizes favor rheological stabilisation. Rheological stabilisation is mostly employed for (viscous) [materials]" [10]. The majority of hollow glass microsphere applications are viscous, leading to rheological stabilization through hindrance of Brownian motion, buoyancy, and frictional forces. The buoyancy force is simply the weight of liquid (or semi-solid) displaced, as shown in Eqn (11.9):

$$F_B = V_s \rho_l g \quad (11.9)$$

where

V_s = the volume of a hollow glass microsphere

ρ_l = density of the liquid phase

g = acceleration due to gravity

The frictional (Stokes) force, also known as the drag force, was first derived to quantitatively explain frictional forces on very small spherical objects (i.e., very small Reynolds number) within viscid media, and is given as

$$F_d = 6\pi\mu rv \tag{11.10}$$

where

μ = matrix viscosity

r = radius of the hollow glass microsphere

v = settling or flotation velocity, depending on density/buoyancy

The gravitational force is simply the mass of an object multiplied by the acceleration due to gravity, and can be rewritten as

$$F_G = V_s\rho_s g \tag{11.11}$$

where ρ_s = density of hollow glass microsphere

Since the gravitational force is equal to the sum of the buoyancy and friction forces, we can rearrange Eqns (11.9)–(11.11) to solve for v, the settling (or flotation) velocity, as follows:

$$F_d = F_G - F_B \tag{11.12}$$

$$6\pi\mu rv = V_s\rho_s g - V_s\rho_l g \tag{11.13}$$

Given the volume of a sphere, Eqn (11.13) reduces to

$$v = \left(2/(9\mu)\right)r^2(\rho_s - \rho_l) \tag{11.14}$$

The negative sign of ($\rho_s - \rho_l$) signifies the natural floatation of glass microspheres; the dispersion difficulties associated with which were detailed in earlier sections. The viscosity in the denominator shows that the greater the viscosity of the matrix, the more inhibited the floating of hollow glass microspheres will be. The square of the radius (with units of meters or feet) indicates that the smaller a particle is—even if it is suspended in a product of low viscosity, or the density difference between materials is very large—the particle will be less likely to settle or float (depending on sign of density difference). Overall, what this equation expresses is, despite initially being technically challenging to incorporate in a homogeneous manner, hollow glass microspheres will tend to stay within the matrix, given ample viscosity and/or density of the matrix.

References

[1] A. Nienow, T. Elson, Untitled, Chem. Eng. Res. Des. 66 (1988) 5.

[2] V. Atiemo-Obeng, W. Penney, P. Armenante, Solid-liquid mixing, in: E. Paul, V. Atiemo-Obeng, S. Kresta (Eds.), Handbook of Industrial Mixing: Science and Practice, John Wiley & Sons, Inc, Hoboken, New Jersey, 2004, pp. 543–545, 547, 564–565.

[3] R. Weetman, Role of the mixing equipment supplier, in: E. Paul, V. Atiemo-Obeng, S. Kresta (Eds.), Handbook of Industrial Mixing: Science and Practice, John Wiley & Sons, Inc, Hoboken, NJ, 2004, p. 1345.

[4] D. Todd, Mixing of highly viscous fluids, polymers, and pastes, in: E. Paul, V. Atiemo-Obeng, S. Kresta (Eds.), Handbook of Industrial Mixing: Science and Practice, John Wiley & Sons, Inc, Hoboken, NJ, 2004, pp. 988–989, 991–992.

[5] E. Fischer, Colloidal Dispersions, Wiley, New York, 1950.

[6] E. Guth, R. Simha, Untersuchungen über die Viskosität von Suspensionen und Lösungen. 3. Über die Viskosität von Kugelsuspensionen, Kolloid Z. 74 (3) (1936) 266.

[7] D.G. Thomas, Transport characteristics of suspension: VIII. A note on the viscosity of Newtonian suspensions of uniform spherical particles, J. Colloid Sci. 20 (3) (1965) 267.

[8] T. Kitano, T. Kataoka, T. Shirota, An empirical equation of the relative viscosity of polymer melts filled with various inorganic fillers, Rheol. Acta 20 (2) (1981) 207–209.

[9] T. Kataoka, T. Kitano, M. Sasahara, K. Nishijima, Viscosity of particle filled polymer melts, Rheol. Acta 17 (2) (1978) 149–155.

[10] Walter Rath, Bodo Muller, Formulating Adhesives and Sealants: Chemistry, Physics and Applications, Vincentz Network, Hannover, 2010, 63.

Index

Note: Page numbers followed by "*f*" and "*t*" indicate figures and tables respectively.

G

H

I

J

Q

R

FSC
www.fsc.org
MIX
Papier | Fördert gute Waldnutzung
FSC® C083411

Kolibri 360 GmbH
Ferdinand-Jühlke-Straße 7
99095 Erfurt, Deutschland
produktsicherheit@kolibri360.de